PHYSICS RESEARCH AND TECHNOLOGY

THEORY OF QUANTA

PHYSICS RESEARCH AND TECHNOLOGY

Additional books and e-books in this series can be found on Nova's website under the Series tab.

PHYSICS RESEARCH AND TECHNOLOGY

THEORY OF QUANTA

MARIAN APOSTOL

nova science publishers
New York

Library of Congress Cataloging-in-Publication Data

ISBN: 978-1-53616-651-4

Published by Nova Science Publishers, Inc. † New York

Contents

1 Introduction

The largest part of this book follows closely the *Quantum Mechanics* (*Non-relativistic Theory*) of Landau and Lifshitz; it includes standard subjects like Schrodinger equation, perturbations, spin, identical particles, quasi-classical theory, scattering theory. The remaining small part is original; it includes the Old Quantum Mechanics, the matricial Quantum Mechanics, a few points in the quasi-classical theory, etc. The main point in writing (and publishing) this book was to give my opinion about the understanding of the Quantum Mechanics. There were many people who did quantum-mechanical calculations, and there are many who still do; some calculations are correct, many others are wrong; only a few understand what they do. They shut up and calculate (as Mermin says); in many cases the quantum-mechanical equations seem to be wiser than people. Along the time, since its inception, the Quantum Mechanics proved itself to be a difficult subject.

At the atomic level (and below) the physical world is completely different from the world at the macroscopic level. The quantum-mechanical particles are not at rest, they do not move along trajectories, the physical quantities are indefinite. At high energies, in scattering experiments, another complication arises; the high-energy scattering implies short times and distances, where the quantum-mechanical energy transfer and the momentum transfer are undefined, so that, imposing a relativistic quantum-mechanical behaviour is conflictual; quantum-mechanical effects are limited in that case to small contributions. Although the relativity can be formally joined with the Quantum Mechanics, through the electron Dirac equation and the S-matrix, the applicability of the Quantum Mechanics at high energies is limited; the quantum field theory and, in particular, the quantum electrodynamics require appreciable limitations in employing the Quantum Mechanics.[1] The quantum-mechanical character of the relativistic fields appears in other, distinct circumstances. The present book deals with the non-relativistic quantum theory.

At the atomic level the measurement process is very different from the measurement process at the macroscopic level. First, since the particles, distances, physical quantities are small and the times are short at this level, the measurement process is indirect; it requires not only a photographic plate or a bubble chamber, but often a complicated chain of electronic equipment; consequently, many of the measurement results at this level are inferred indirectly from acquired data; but the inference requires a theoretical framework, and the Quantum Mechanics had to provide an "interpretation" of the measuring data. Second, the measurement process at the atomic level is statistical, *i.e.* the result of the measured quantity is a mean value, which implies a probability for various values of that physical quantity. The quantum-mechanical motion is statistical. Finally, since the atomic world is such a "delicate" world, any measuring process disturbs the physical object on which the measuring process is exerted; which means that simultaneous measurements of different quantities may not be possible. All these features are very far from what we are used with at the macroscopic level, and the Quantum Mechanics had to devise new concepts and to coin new words in order to describe such situations. Being so distant from our usual experience and our intuition, acquired at the macroscopic level, the quantum-mechanical concepts and words have generated, and still do nowadays, many confusions, obscure descriptions and formulations, very often an outrightly wrong prose, paroxistic attitudes or even desperation. "Nobody understands Quantum Mechanics" (Feynman), "If the Quantum Mechanics hasn't profoundly shocked you, you haven't understood it yet" (Bohr, approximate quotes). I hope I succeeded in keeping at a minimum the errors and the misleading formulations in the exposition given in the present book. In presenting the Quantum Mechanics a complementarity principle works: if you try to be very accurate and complete, you end in error; if you try to avoid any error, you are forced to si-

[1]L. Landau, "On the Quantum Theory of Fields", in *Niels Bohr and the Development of Physics*, ed. W. Pauli, Pergamon Press (1955) and references therein; W. Heitler, *The Quantum Theory of Radiation*, Oxford University Press (1944).

lence. Perhaps more than in any other branch of Physics, any problem of Quantum Mechanics should be solved within its experimental context, with a clear understanding of the experimental conditions and the measured quantities.

The first key point in constructing the Quantum Mechanics was provided by the existence of the discrete energy levels, through Planck's hypothesis, Bohr's atomic states and Einstein's quanta of energy; hence, the notion of periodic (or quasi-periodic) orbits appeared, the Bohr-Sommerfeld quantization rules and the whole theory of what we call today the "quasi-classical approximation"; this was in fact the Quantum Mechanics, usually called the Old Quantum Mechanics; though it talks about particles, the notion of wave was implicit in this construction. The Old Quantum Mechanics was connected to the classical concepts of the Mechanics by Heisenberg in his matricial Quantum Mechanics, which led to matrices for the physical quantities (*i.e.* undefined physical quantities), probability for the measured quantities, the uncertainty principle and the complementarity notion. The second key point was de Broglie's wavelength associated to the quantum-mechanical motion, which led to Schrodinger's wave (undulatory) mechanics (Dirac contribution, which is mentioned often in this context, was on the less-important formalization side). It turned out that the two forms of Quantum Mechanics, matricial and undulatory, are in fact two languages for the same situation.

Along these lines, the Quantum Mechanics was developed into a working formalism. However, the Schrodinger wavefunction (which prompted a more profound understanding of the quasi-classical motion) raised a very troublesome problem, which, very likely, is the main cause which hampers the understanding of the Quantum Mechanics. The problem is: what are the quantum-mechanical objects, particles or waves? This is the particle-wave duality question, which is probably the main source of the "shocking" character of the Quantum Mechanics. The answer to this question resides in the very Schrodinger equation. Suppose that we wish to make an equation about the energy of a moving quantum-mechanical particle, by using a wave ψ; then, according to the quantization rules, we should use $i\hbar\partial/\partial t$ for energy and $-i\hbar grad$ for momentum, where $\hbar$ is Planck's constant; the mechanical energy for the motion of a particle with mass m in the potential V

gives then the Schrodinger equation $i\hbar\partial\psi/\partial t = -(\hbar^2/2m)\Delta\psi + V\psi$. Except for the wavefunction ψ, the Schrodinger equation includes quantities which are specific to a particle: mass, energy, momentum, potential, as well as the classical mechanical form of the energy. Therefore, in using the Schrodinger equation it seems that we are talking about a particle; but what possibly means this ψ for a particle?. In what sense is ψ an attribute of a particle? Is it a measurable quantity? According to Heisenberg's conception, we may view the position $\mathbf{r}$ as a set of states of the particle, and, equally well, we may view the ψ as a set of states too (indeed, the Schrodinger equation can easily be viewed as an eigenvalues and eigenfunctions equation). Then, one state in a set may be written as a linear combination of the states in the other set; the coefficients of the combination are scalar products of states; they are functions of the form, say, $\psi(\mathbf{r})$, omitting the time (or, equally well, $\mathbf{r}(\psi)$). We recognize here Dirac's notation. Consequently, $\psi(t, \mathbf{r})$ (we re-introduce the time as an argument) gives the probability of the $\mathbf{r}$-state in the ψ-state, through $\mid \psi \mid^2$, as we know from Heisenberg's original theory. It follows that the amplitude of ψ is the square root of a probability. This is the core of the disturbing problem, because it raises the question: does a particle have a probability? The difficulty in answering this question resides in a common confusion. We are used to say that a particle has mass, charge, spin, etc and also it has a position, an energy, a momentum, etc. The first class of quantities are viewed as being intrinsic, and they define the particle; but the second class of quantities belong to the motion of the particle. Now, we should recall that mass may be attributed to a particle through motion mechanisms, like the curved space-time in gravitation, or the Higgs mechanism in quantum field theories; the spin arises from the relativistic motion, the charge may be given by interaction like in the renormalization of the quantum electrodynamics; it seems that all the "intrinsic" attributes of a particle arise from its motion, like those in the second class above. The particle in itself remains an undefined object, so that the question "what is this? a particle or something else?" is illegitimate, since one term of this question is not definite. On the other hand, we should notice that all the above quantities exist insofar as they are measurable, their definition being the very experimental context by which they are measured; of course, there exist in Physics mathematical quantities which are defined by their relation with others, but even in mathematics we need to view some quantities as primary quantities, at least conventionally.

From the discussion given above we may say that ψ is a perfectly legitimate, or at least relevant, attribute of a particle, because the probability $| \psi |^2$ is a measurable quantity, and the amplitude A in $\psi = Ae^{i\Phi}$ is measurable. The Quantum Mechanics introduces a new attribute for the particles, arising from their quantum-mechanical motion, namely the probability $| \psi |^2$. The question: "what are these objects, particles or waves?" becomes meaningless, because it tries to compare undefined objects with their attributes; it is precisely the attributes which define an object, called particle, so we may answer this question: this object is a particle with probabilities (*i.e.*, a particle which moves with probabilities). The object we are talking about is a "particle with wave attributes" (or a "wave with particle attributes").

The discussion above is about a particle with mass (possibly charge, etc) to illustrate its new attribute, the probability. But equally well we may think of a photon, or other massless quantum fields; as well as other, non-relativistic, quantum fields (with their own equations of motion). The quantum-mechanical probability is preserved, though it manifests itself in different conditions. The present book deals with quantum particles, not fields.

The confusion embedded in the "duality dilemma" is favoured by the fact that the probabilities are represented by a wave, a probability wave. In this respect, while the amplitude can be written as $A = \sqrt{| \psi |^2}$, it is the phase which raises the question of whether it is measurable or not. It is the phase which is responsible for the interference arising from the superposition of the wavefunctions (the superposition implied by the linearity of the Schrodinger equation); the genuine undulatory character of the probability wave arises from its phase. In this respect (leaving aside the inter-connection between phase and amplitude in the Schrodinger equation), we should recall that the measurable parity is given by the phase of the wavefunction, the measurable symmetry of the wavefunction for identical particles is generated by the phase of the wavefunction, the phase factors associated with rotation symmetries generate linear combinations which lead to measurable effects, and, not in the least, the electromagnetic phase in the Aharanov-Bohm effect is measurable and the gauge interactions generated by phase factors are also measurable; the phase of the quasi-classical wavefunction is the measurable mechanical action; the phase determines the coherence, which is quite measurable. We may admit that the phase of the quantum-mechanical wavefunction is a measurable quantity, at least partially. The quantum-mechanical probability wavefunction may be viewed as a definite physical quantity (*i.e.*, measurable), generated by the quantum-mechanical motion, which is an attribute of quantum-mechanical objects like particles or fields, participating in the definition of these objects. The discussion can be prolonged in directions which become too philosophical to be of any interest.

Having elucidated the meaning of the Quantum Mechanics, it remains to briefly discuss a few worthy, related points. The first is the relation of the Quantum Mechanics to relativistic quantum field theories, in particular the quantum electrodynamics, at high energies, which has been briefly touched at the beginning of this chapter. The second point refers to the condensed matter. The quantum-mechanical states which are specific to condensed matter (macroscopic bodies) are extremely dense (except, of course, quantum-mechanical states like those of magnetic moments for instance, which, however, do not characterize the condensed matter). The quantum-mechanical numbers, like energy, and quantum-mechanical wavefunctions are not defined in condensed matter. Moreover, the macroscopic bodies are not isolated from their environment, such that their quantum-mechanical behaviour is described by a density matrix. At thermal equilibrium, this density matrix is the Gibbs distribution, and is called statistical matrix, or statistical distribution; the thermal properties imply summations of diagonal elements over quantum-mechanical states; therefore, in these summations free-particle states may be used. From all the Quantum Mechanics there remains in this case only the quantization rules and the state dimension $2\pi\hbar$ per degree of freedom in the phase space. These three ingredients suffice to account for quantum-mechanical statistical effects, as in ideal quantum gases, for instance, or in the black-body radiation; the thermal properties of the condensed matter arise from the quantum-mechanical statistics of their elementary excitations. In this context, we should re-iterate here that quantum-mechanical effects may appear at the macroscopic scale, like superfluidity, superconductivity, laser coherence; such effects appear through the quantum-mechanical correlated motion at the atomic level of the constituents of the condensed bodies.

There exists a problem which is not usually included in the "established" body of the Quantum Mechanics; this is the coherence problem. In general, the quantum-mechanical wavefunction depends on the initial conditions (position, time, etc) of the particle; this dependence is accounted by a constant phase factor; when superposing wavefunctions of different particles, these phase factors are included in the superposition; they may produce interference, when the probability is computed; usually these phase factors are randomly distributed, such that they do not generate a "constructive" interference, and the effect of the superposition is proportional to the number of particles; on the contrary, when these phase factors are equal, the constructive interference appears and the effect is proportional to the squared number of particles; this is the coherence phenomenon; when coherent, the wavefunction corresponds to the whole assembly of particles, as if the particles would be correlated. It may happen that the particles exchange very small amounts of energy; consequently, they do that in a very long time; therefore, there is sufficient time for them to behave as an entangled assembly of quantum-mechanical particles; in such cases they become coherent. This happens for superconductivity, superfluidity, lasers, etc. It may also happen for neutrinos. In view of the very small scattering amplitude of the neutrinos when interacting with matter, they transmit to the material particles small amounts of energy in a long time, such that the latter behave as a coherent assembly; the scattering amplitude becomes then very large.[2]

[2] J. Weber, "Method for observation of neutrinos and antineutrinos", Phys. Rev. **C31** 1468 (1985); "Apparent observation of abnormally large coherent scattering cross sections using keV and MeV range antineutrinos, and solar neutrinos", Phys. Rev. **D38** 32 (1988); M. Apostol, "Weber's coherent scattering and neutrino detection", Roum. Reps. Phys. **60** 315 (2008).

2 Beginnings

2.1 Introduction

About 1900 there was a consensus among phycisists that matter is composed of atoms, which may unite and form molecules. There were about 90 species of atoms, corresponding to chemical elements. Any atom consists of a very small central nucleus, with a positive electric charge, surrounded by a neutralizing cloud of negatively-charged electrons. The whole mass of the atom is concentrated practically in its nucleus. Later on, it was discovered that the atomic nucleus includes positively-charged protons and neutral neutrons; these were called collectively nucleons. The atomic nucleus contains Z protons, A nucleons and $A - Z$ neutrons, where Z is the atomic number of the chemical element in the Mendeleev table and A is the mass number. The electrons and the nucleons were viewed as particles; the electrons were thought as practically point-like particles, while the nucleons were viewed as having finite, very small, dimensions. This is the "discrete" structure of matter and the atomistic conception.

At that time, the general conception of motion of the particles was that offerred by the classical mechanics of Newton and by the statistical mechanics, the latter embodied in the kinetic-molecular theory. The classical mechanics was extended to the motion of the continuous matter, described by the theories of elasticity and fluids. There was also known the theory of electricity and magnetism, both in vacuum and matter, which describes the electromagnetic interaction. It was known that the electromagnetic radiation consists of waves which propagate in vacuum with velocity $c = 3 \times 10^{10} cm/s$, probably a universal constant, independent of the reference frame.[1]

[1] All the figures relating here to physical quantities should be taken approximately, as being affected by small errors.

Gradually, we arrived at knowing that the electron mass is $m = 10^{-27}g$ and its charge $e = -4.8 \times 10^{-10}esu$ $(statcoulomb)$; the nucleon mass (approximately the same for proton and neutron) is $\simeq 2000$ electron mass; the proton charge is equal to minus the electron charge; the nucleon dimension is $\simeq 10^{-13}cm$. The electromagnetism seems to impose a limit of the order $e^2/mc^2 = 3 \times 10^{-13}cm$ for the electron size (called the classical electromagnetic radius).

An amount of any substance numerically equal with its molecular mass in grams is called mol; any mol contains the Avogadro number 6×10^{23} molecules. From the law of the ideal gases we may infer that the mean separation distance between gas molecules in normal conditions of pressure (atmospheric pressure $10^6 dyn/cm^2$) and temperature (room temperature $300K$, $1K = 1.38 \times 10^{-16}erg$) is of the order $35 \times 10^{-8}cm$ ($10^{-8}cm$ is called angstrom, Å). Therefore, we may infer that in liquids or solids the mean inter-molecular spacing is of the order of a few Å; this is the order of atomic dimensions.

2.2 Planck's quantum of mechanical action

Electric charges in oscillatory motion absorb and emit electromagnetic radiation. The electrons in atoms may do that. We may conceive of an isolated portion of matter containing electromagnetic radiation in a volume V, without receiving nor releasing radiation from or to outside. Therefore, we may call such an assembly a black body. It is conceivable that a black body may be in thermal equilibrium, at some temperature T. What would then be the energy distribution $U(\nu)d\nu$ of the radiation with respect to frequencies ν? We know that the electromagnetic radiation is a superposition of plane waves with frequency ν and wavevector $\mathbf{k}$, such that $\nu = ck/2\pi$. The black-body radiation should have all frequencies ν from zero to infinity and all the wavevectors $\mathbf{k}$ (except $\mathbf{k} = 0$). In the plane wave we have a phase $\mathbf{kr}$, and a state of plane wave is characterized by $\Delta k_i L_i = 2\pi$, where $i = x, y, z$ denotes the three cartesian coordinates and L_i denotes the length along the i-th coordinate; $L_x L_y L_z = V$. Therefore, the wavevector components are given by $k_i = (2\pi/L_i)n_i$, where n_i is any integer. For a macroscopic sample, the shape does not matter. The number of states comprised between k and $k + dk$ is $dn_x dn_y dn_z =$

$4\pi n^2 \Delta n = \frac{4\pi V}{(2\pi)^3} k^2 dk$, or $(4\pi\nu^2/c^3)d\nu$; here, we must insert a factor 2 arising from the two polarizations of the radiation; therefore, the number of states of plane waves is $(8\pi\nu^2/c^3)d\nu$.

It is also known that any electromagnetic plane wave has an amplitude which moves like a harmonic oscillator with mass unity and frequency ν; this arises from Maxwell equations. This is not an oscillator in the direct space, but one whose coordinate is the Fourier transform of the electromagnetic vector potential. The amplitude is a vector transverse to the propagation direction, so it has two (polarization) components. The statistical distribution of a harmonic oscillator is $\beta e^{-\beta E} dE$, where E is its energy (varying from zero to infinity) and $\beta = 1/T$ is the inverse of the temperature T; it follows that the mean energy of the oscillator is $1/\beta = T$, so that the frequency distribution of the black-body radiation would be $U(\nu)d\nu = (8\pi\nu^2 T/c^3)d\nu$. This is known as the Rayleigh-Jeans law; it is verified experimentally for low frequencies, but not for high frequencies, where it gives an infinite distribution (known as the ultraviolet "catastrophe"). We should say that the frequency distribution of the black-body radiation can be measured by taking the radiation coming from the black body through a small hole, dispersing it through a prism and measuring the intensity on a photographic plate. The experimental distribution for high frequencies is given by Wien's law $U(\nu) \sim \nu^3 e^{-h\nu/T}$, where the experimental fit gives

$$h = 6.6 \times 10^{-27} erg \cdot s \ . \tag{2.1}$$

The understanding of this law was lacking, until Planck assumed in 1900 that the oscillator energy is[2]

$$E = nh\nu \ , \tag{2.2}$$

where n is any positive integer (zero included). Then, the statistical distribution is computed as

$$Ce^{-\beta nh\nu} \ , \ C\sum_{n=0}^{\infty} e^{-\beta nh\nu} = \frac{C}{1-e^{-\beta h\nu}} = 1 \ , \tag{2.3}$$

[2]M. Planck, "Zur Theorie des Gesetzes der Energieverteilung im Normalspektrum", Verhandl. d. Deutchen physikalischen Gesellschaft **2** 237 (1900).

and the mean energy of the oscillator is

$$\overline{E} = \overline{n}h\nu = C\sum_{n=0}^{\infty} nh\nu e^{-\beta nh\nu} = \frac{h\nu}{e^{\beta h\nu} - 1} ; \tag{2.4}$$

hence

$$U(\nu) = \frac{8\pi h\nu^3}{c^3}\frac{1}{e^{\beta h\nu} - 1} , \tag{2.5}$$

which gives immediately the Rayleigh-Jeans and Wien limiting cases. This equation is well verified experimentally.

According to Planck's assumption, the oscillator has discrete energy levels, the energy being exchanged in integers of the quantum of energy $\varepsilon = h\nu$; as if there were n quanta for each energy levels, $\overline{n}$ being the mean occupation number of quanta of the n-th energy level. The energy levels define the states of the oscillator, the lowest in energy being the ground-state or the fundamental state (level); those with higher energies are excited states. These states have well-defined energies. This assumption remained for a long time misterious, and proved finally to have far reaching implications. The Planck constant h is the quantum of mechanical action. Obviously, processes of exchanging energy in quanta call for a new mechanical theory, different from Newton's mechanics, which may be called quantum mechanics. The new mechanical theory should act at the atomistic level and should pass into the Newton's theory at the macroscopic level (though, of course, exceptions may exist). We note that for $h \rightarrow 0$ the number n goes to infinity and the energy levels are continuous, as in Newton's mechanics.

2.3 Photoelectric effect

At the end of the 19th century there was known the photoelectric effect. The effect consists of liberating electrons from metallic surfaces when illuminated by light, or ultraviolet radiation. The electrons can be identified by their motion in electric and magnetic fields, and their energy can be measured by stopping them in a voltage. The curious fact was that the electron energy does not depend on light intensity, but instead it is given by

$$E = h\nu - W \quad , \tag{2.6}$$

where ν is the light frequency and W a constant specific to the metallic surface. The intensity of the photoelectric current is proportional to the light intensity. Einstein noticed from experimental data that the constant h in equation (2.6) is the same as Planck's constant. In accordance with Planck's hypothesis he advanced the following explanation for the photoelectric effect:[3] each electron in metal is struck by a light corpuscle, which gives energy $h\nu$ to the electron; thereafter, the electron spends an amount W of this energy to escape from the metal, and the remaining part E is the energy we measure. W is the work function of the metal. Light is made by many such corpuscles, for each frequency ν, each of them having a chance to transfer energy $h\nu$ to an electron on striking it, explaining thus the dependence of the photoelectric effect on the light intensity; the latter being proportional to the number of light corpuscles. Later on, it was observed, indeed, that very weak light thrown upon a photographic plate gives small dots, in those points where light is absorbed by atoms in the plate; and, on increasing the light intensity, the spots become a continuous patch. The light corpuscle was later called photon.

Einstein's photon concept

$$\varepsilon = h\nu \tag{2.7}$$

implies a discrete exchange of energy in quanta $h\nu$, in accordance with Planck's hypothesis, as well as a probabilistic nature of the interaction process: not all light quanta are absorbed, nor all electrons ejected from metal, but there is a certain temporal rate of the process. The photon corresponds to the constituent quantum in Planck's energy law $E = nh\nu$, as if the radiation oscillator is made out of many photons (any integer).

Nevertheless, it is extremely difficult to reconcile the corpuscular nature of the light proposed by Einstein with the undulatory nature of the light, known very well from electromagnetism, and widely verified experimentally.

[3] A. Einstein, "Uber einen die Erzeugung und Verwandlung des Lichtes betreffenden heuristischen Gesichtspunkt", Ann. Physik **17** 132 (1905).

2.4 Double-slit experiment

With the technical means of today we can produce very weak light, *i.e.* electromagnetic radiation with very weak intensity. If we try to measure the electric and magnetic fields in such a light ray, we get not definite results; but letting the ray to fall on a photographic plate we get a few small spots, which reasonably measure the light intensity and indicate the atoms where the light is absorbed. The position of these spots cannot be specified, but on keeping illuminating the plate, after a while, shorter or longer, the spots become a continuous patch which is the shadow of the hole through which we have passed the light and prepared the light ray. More, we can get an interaction of the ray with an electron beam, and see that the interaction is produced in points, randomly distributed along the ray. This indicates that the ray seems to be continuous, but suddenly, in the interaction process it becomes localized on atomistic distances, either on its ray course, or on the photographic plate. (This was latter called the "collapse" of the wavefunction). The interaction process of the electromagnetic radiation with matter seems to have a probabilistic nature. Moreover, we may pass a light ray with a given frequency through two slits pierced in a shield, with dimensions comparable with radiation wavelength and not far away from one another. We may arrange to have a very weak light. Then, we will see one point on the photographic plate, thereafter another one, and so on, at varius positions without any special pattern of localization at first sight. However, after a while, we may notice that the points on the plate become in fact arranged in an interference pattern, according to the wave nature of light. That means that even a photon is a wave, which passes in fact through the two slits, as any wave, at once. A wave is delocalized in space and time, and, still, the interaction of a light wave with matter appears probabilistically, at any point and moment, and localized on atomistic distances.

This is the general qualitative picture of the quantum-mechanical phenomena.

This dualistic behaviour, *i.e.* the duality corpuscle-wave, turns out to be the general nature of matter and forces at the quantum level, *i.e.* at least as long as the dimensions are of the order of the atomistic dimensions at most, the energies are of the order of atomistic energies at most and the mechanical action changes by amounts comparable with Planck's constant h, which is the quantum of mechanical action. We use to say that the quantum level is the level of atomistic phenomena. We note that the time is a parameter in the quantum-mechanical phenomena at non-relativistic velocities. Later on, it was noted that the quantum nature is manifest also in some macroscopic phenomena.

2.5 Specific heat of solids

In 1907 Einstein[4] extended Planck's hypothesis from radiation oscillators to material oscillators. The atoms in solids may oscillate about equilibrium, fixed positions; therefore, their mean energy should be given by the radiation distribution function, integrated over frequencies, such as to match the number of normal modes of oscillation to the number of degrees of freedom. At low temperature the mean energy behaves like the radiation distribution at high frequencies; then, the Wien law gives the mean energy $\overline{E} \sim T^4$ and the specific heat of solids $\sim T^3$, a law verified experimentally (and partly known at that time). At high temperature, the mean energy goes like the Rayleigh-Jeans law $\overline{E} \sim T$, which leads to the experimentally well-known Dulong-Petit law of constant specific heats of solids. The extension given by Einstein strenghtened the idea that Planck's hypothesis may have a fairly general character. The elastic vibrations of the solids and the atom oscillations in solids were given a quantum-mechanical treatment in accordance with Planck's hypothesis by Debye in 1912;[5] the quantum-mechanical corpuscle of elastic vibrations (*i.e.* of elastic interaction forces) was called phonon, and the $\sim T^3$-law of specific heat at low temperature is called the Debye law.

[4] A. Einstein, "Die Plancksche Theorie der Strahlung und die Theorie der spezifischen Warme", Ann. Physik **22** 180 (1907).

[5] P. Debye, "Zur Theorie der spezifischen Warme", Ann. Physik **39** 789 (1912).

2.6 Atomic energy levels

At the end of the 19th century it was known that gases illuminated by light exhibit spectral absorption lines at wavelengths λ; for hydrogen these wavelengths are given by

$$\frac{1}{\lambda} \sim \frac{1}{n^2} - \frac{1}{m^2} \ , \tag{2.8}$$

where n, m are non-zero integers ($n < m$); the absorbed light is re-emitted in all directions. It follows that atoms absorb and emit electromagnetic radiation according to a law governed by integers. Since $\lambda = c\nu$ and energy is $E = h\nu$ according to Planck and Einstein, it follows that the hydrogen atom has energies of the form

$$E_n \sim -\frac{1}{n^2} \ , \tag{2.9}$$

the minus sign accounting for the fact that these energies correspond to bound states; and the hydrogen atom exchanges energies of the form

$$E_{nm} \sim \frac{1}{n^2} - \frac{1}{m^2} \ , \tag{2.10}$$

by absorbing or emitting corresponding photons. The electrons moving about the atomic nucleus have an accelerated motion, and, therefore, according to the electromagnetic theory, they should emit energy continuously, until they would fall upon the nucleus; such way, the atom would be unstable. The hypothesis of energy quanta would tell that this is not the case, the atoms being in stationary states, with definite energies E_n, and they may pass (jump) from one state to another, changing their electron configuration, absorbing or emitting photons. This hypothesis of discrete, stationary atomic energy levels, corresponding to quantum-mechanical states, was put forward in 1923 by Bohr in a series of three papers.[6] Bohr's hypothesis implies that electrons have not a definite trajectory, as Newton's mechanics would predict. The energies given by equation (2.9) indicate a ground-state (fundamental state) corresponding to $n = 1$ and a continuum of

[6]N. Bohr, "On the constitution of atoms and molecules", Phil. Mag. **26** 1, 476, 857 (1913).

states for high $n \gg 1$, corresponding to a picture in accordance with the classical mechanics of Newton. This circumstance is called Bohr's correspondence principle;[7] in accordance with Planck's hypothesis, the above equations should include h^2n^2, and for $n \rightarrow \infty$, $h \rightarrow 0$.

2.7 Franck-Hertz experiment

Franck and Hertz passed an electron beam through a gas and measured the energies of the electrons which sufferred collisions with the gas atoms; they found that the electrons loss energy in finite amounts, which indicate that the atoms are excited on superior energy levels by electron collisions.[8] This is a direct proof of Bohr's hypothesis. Of course, atom ionization occurs also in this experiment, where the electrons are liberated from atom.

2.8 Compton effect

According to the electromagnetic theory, if an electromagnetic radiation hits an electric charge, it is scatterred off but does not change frequency. In 1923 Compton scatterred X rays off paraffin, and observed that the scattering rays change frequency.[9] The change of frequency can be measured by passing X-rays through a crystal, where they produce an interference pattern. If the radiation wavevector changes by an angle, say, θ, then the change would be $\Delta k = k(1 - \cos\theta)$, or $\Delta k = 2k \sin^2 \theta/2$ for small θ; that means that the wavelength changes by $\Delta\lambda = 2\lambda \sin^2 \theta/2$. This was not the result observed experimentally by Compton.

Since the radiation changes frequency, it follows, according to the hypothesis of the energy quanta, that the radiation, *i.e.* the photon, changes its energy; therefore, the photon, as any other corpuscle,

---

[7] N. Bohr, "Uber die Serienspektra der Elemente", Z. Phys. **3** 423 (1920).

[8] J. Franck and G. Hertz, "Uber Zusammenstosse zwischen Elektronen und Molekulen des Quecksilberdampfes und die Ionisierungsspannung desselben", Verhandl. d. Deutchen physikalischen Gesellschaft **16** 457 (1914).

[9] A. H. Compton, "A quantum theory of the scattering of X-rays by light elements", Phys. Rev. **21** 483 (1923).

should also change momentum; therefore, the photon should have a momentum. If the mechanical law $\mathbf{v}=\partial\varepsilon/\partial\mathbf{p}$ is valid, for photon it gives

$$\varepsilon = cp \quad , \tag{2.11}$$

hence

$$p = \frac{\varepsilon}{c} = \frac{h\nu}{c} = \frac{h}{\lambda} \tag{2.12}$$

and $\mathbf{p} = h\mathbf{k}/2\pi$, where $\mathbf{p}$ is the photon momentum, $\mathbf{k}$ is the radiation wavevector and $\mathbf{v}$ denotes the velocity (for light $v = c$). Since the wavevector is such that $k_i = (2\pi/L_i)n_i$, where n_i is any (non-zero) integer and L_i is the length along the i-th coordinate, it follows that the i-th component of the momentum is quantized according to $p_i = hk_i/2\pi = hn_i/L_i$; a relationship which implies integers, similar with the energy quantization $\varepsilon = nh\nu$. This relationship was called later the quantization of the momentum.

Now, we may write down the momentum and energy conservation in the photon-electron scattering

$$\mathbf{p} = \mathbf{p}' + \mathbf{P} \ , \ , pc = p'c + P^2/2m \quad , \tag{2.13}$$

where $\mathbf{p}$ is the photon momentum before collision, $\mathbf{p}'$ is the photon momentum after collision and $\mathbf{P}$ is the electron momentum; initially, the electron is assumed to be at rest; m denotes the electron mass. It follows immediately the wavelength change

$$\Delta\lambda = 2\frac{h}{mc}\sin^2\frac{\theta}{2} \ ; \tag{2.14}$$

this relationship was what Compton saw in his experiment. $h/mc \simeq 2.2 \times 10^{-10} cm$ is called Compton's wavelength.

Compton's hypothesis of a photon momentum $p = h/\lambda$ substantiated appreciably Einstein's photon hypothesis.

2.9 Particles are waves

If the electromagnetic radiation consists of photon particles with energy $\varepsilon = h\nu$ and momentum $p = h/\lambda$ and if the atoms and material particles like electrons have energies which are multiple of the energy quantum and the energy exchange between radiation and material particles proceeds in energy quanta, why the momenta of the material particles would not be of the form $p = h/\lambda$, such that the momentum exchange between radiation and material particles be consistent with the radiation momentum? This would mean that material particles have a wavelength, *i.e.* they would be waves. This hypothesis was advanced in 1924 by de Broglie.[10] It follows that we should associate a wavefunction to particles, both material and radiation particles (photons); and, of course, to view waves as particles in their interaction processes.

2.10 Electron diffraction

Between 1925-1927 Davisson and Germer fired electrons accelerated at energies $\sim 50V$ at the surface of metallic nickel; they observed an interference pattern at angles θ, given by the diffraction law $2d \sin\theta = n\lambda$, where d is the distance between the crystalline planes of nickel atoms, λ is the wavelength of the electron waves and n is any (small) integer.[11] This experiment confirmed directly de Broglie hypothesis of material waves. From X-ray diffraction the distance $d \simeq 1$Å was known; angle θ was found $65°$; for $n = 1$ we get $\lambda = 1.7$Å. The energy $50eV$ is $\simeq 8 \times 10^{-11} erg$ (electron charge is $-e = 1.6 \times 10^{-19} C$, $1eV = 1.6 \times 10^{-12} erg$); it follows the momentum $p \simeq 4 \times 10^{-19} g \cdot cm/s$ and the wavelength $\lambda = h/p = 1.5$Å, which agrees well with the experimental value.

2.11 Stern-Gerlach experiment. Space quantization

In 1922 there appeared another piece of evidence for discrete states of motion. Stern and Gerlach passed a beam of silver atoms through an

[10] L. de Broglie, "A tentative theory of light quanta", Phil. Mag. **47** 446 (1924).
[11] C. J. Davisson and L. H. Germer, "Reflection of electrons by a crystal of nickel", Proc. Nat. Acad. Sci. USA **14** 317 (1928).

inhomogeneous magnetic field. In a magnetic field $\mathbf{H}$ a particle with magnetic moment $\boldsymbol{\mu}$ acquires an energy $-\boldsymbol{\mu}\mathbf{H}$. The particles with energy $\boldsymbol{\mu}\mathbf{H} > 0$ will be deflected towards higher H, those with energy $\boldsymbol{\mu}\mathbf{H} < 0$ will be deflected towards lower H, in order to minimize the energy. If the directions of the magnetic moments are continuously distributed, a continuous patch will appear on the detector screen. On the contrary, Stern and Gerlach obtained two distinct spots.[12] It means that the magnetic moment of silver atoms in a magnetic field has only two distinct orientations. The magnetic moment is related to the angular moment, so the angular moment has discrete orientations in space, it has distinct states. Other atoms, with other magnetic and angular moments, may have several orientations. This was called space quantization.

[12]W. Gerlach and O. Stern, "Der experimentelle Nachweiss des magnetischen Moments des Silberatoms", Z. Phys. **8** 110 (1922); "Der experimentelle Nachweiss des Richtungsquantelung im Magnetfeld", Z. Phys. **9** 349 (1922); "Das magnetische Moment des Silberatoms", Z. Phys. **9** 353 (1922).

3 Newton's Classical Mechanics

3.1 Lagrange's theory

Newton's mechanics describes the mechanical motion by coordinates q_α, $\alpha = 1, 2, ...s$, where s is the number of degrees of freedom; s may be finite or infinite and the distribution of coordinates may be discrete or continuous. The mechanical state is defined by coordinates q and velocities $\dot{q} = dq/dt$, where t denotes the time and q is a generic notation for q_α. The problem is to find the trajectory $q(t)$. The coordinates q may be the coordinates of the position $\mathbf{r}$ of a particle, not necessarily cartesian; $\mathbf{v} = \dot{\mathbf{r}}$ is the velocity.

The stationary state of the mechanical action

$$S = \int_{t_0}^{t} dt' L(q, \dot{q}, t') \tag{3.1}$$

gives the Lagrange equations of motion

$$\frac{d}{dt}\frac{\partial L}{\partial \dot{q}} - \frac{\partial L}{\partial q} = 0 \tag{3.2}$$

which solve the problem. The function L is the lagrangian, or the Lagrange's function; it includes a term which is quadratic in velocities ("kinetic energy") and a "potential-energy" function, such that the solution of the Lagrange equations be possible in the terms stated above. The typical example is the lagrangian

$$L = \frac{1}{2}m\dot{q}^2 - U(q, t) \tag{3.3}$$

of a particle with mass m, where U is the potential function. Equation (3.2) is Newton's law of motion with momentum $p = \partial L/\partial \dot{q}$ and force

$\partial L/\partial q$. The solution depends on the initial condition q_0 and $\dot{q}_0$ at momentum t_0.

The equations of motion have $2s$ integration constants; they are integrals of the motion. If the time is uniform, *i.e.* if L does not depend on time,

$$\frac{d}{dt}L = \frac{\partial L}{\partial q}\dot{q} + \frac{\partial L}{\partial \dot{q}}d\ddot{q} = \dot{q}\frac{d}{dt}\frac{\partial L}{\partial \dot{q}} + \dot{\frac{\partial L}{\partial \dot{q}}}d\ddot{q} = \\ = \frac{d}{dt}\left(\dot{q}\frac{\partial L}{\partial \dot{q}}\right) \quad , \tag{3.4}$$

hence energy

$$E = \dot{q}\frac{\partial L}{\partial \dot{q}} - L \tag{3.5}$$

is conserved (*i.e.* it is a constant along real trajectories); the motion is said to be conservative, as for a closed motion. For a particle with the lagrangian given by equation (3.3) the energy is

$$E = \frac{1}{2m}p^2 + U \quad , \tag{3.6}$$

where we use the momentum $p = m\dot{q}$. In equation (3.4) and in all similar equations the summation over label α is included.

If the space is uniform (homogeneous), for a closed motion of several particles, the variation of the lagrangian is

$$\delta L = \sum_a \frac{\partial L}{\partial \mathbf{r}_a}\delta\mathbf{r} = 0 \tag{3.7}$$

for a variation $\delta\mathbf{r}$ of all the positions $\mathbf{r}_a$, where a labels the particles; it follows that the total momentum

$$\mathbf{P} = \sum_a \frac{\partial L}{\partial \mathbf{r}_a} \tag{3.8}$$

is conserved; in particular this holds for particles interacting through pair-wise potential functions $U(\mathbf{r}_a - \mathbf{r}_{a'})$; mutual forces $\partial U/\partial\mathbf{r}_a = -\partial U/\partial\mathbf{r}_{a'}$ are equal and opposite in this case, which is Newton's principle of action and reaction.

If the two- or three-dimensional space with one particle is isotropic, *i.e.* if the lagrangian is invariant to rotations $\delta\mathbf{r} = \delta\boldsymbol{\alpha} \times \mathbf{r}$, where $\delta\boldsymbol{\alpha}$ is the infinitesimal rotation angle and $\mathbf{r}$ denotes the position,

$$\begin{aligned} \delta L = \frac{\partial L}{\partial \mathbf{r}}\delta\mathbf{r} + \frac{\partial L}{\partial \mathbf{v}}\delta\mathbf{v} = \dot{\mathbf{p}}\delta\boldsymbol{\alpha} \times \mathbf{r} + \mathbf{p}\delta\boldsymbol{\alpha} \times \mathbf{v} = \\ = \frac{d}{dt}(\mathbf{r} \times \mathbf{p})\delta\boldsymbol{\alpha} \quad , \end{aligned} \tag{3.9}$$

hence the conservation of the angular momentum

$$\mathbf{M} = \mathbf{r} \times \mathbf{p} \ , \ \ M_i = \varepsilon_{ijk} x_j p_k \quad , \tag{3.10}$$

where ε_{ijk} is the antisymmetric tensor of rank three and $x_i\,, p_i$ are the Cartesian components of the position and momentum, respectively. These integrals of the motion are additive. There are not other additive integrals of the motion.

The integrals of the motion may be useful for solving the Lagrange equations, though there are cases where they are not. For instance, for three or more interacting particles there is no one-particle energy; in the same case with Newton's law of gravitational interaction not all of the global integrals of motion are algebraic, which again precludes one-particle solutions; the same problem exhibits chaotic dependence on the inital conditions;[1] such a problem may be termed non-integrable.

3.2 Hamilton's theory

The change in the lagrangian is

$$\begin{aligned} dL = \frac{\partial L}{\partial q}dq + \frac{\partial L}{\partial \dot{q}}d\dot{q} = \dot{p}dq + pd\dot{q} = \\ = d(p\dot{q}) + \dot{p}dq - \dot{q}dp \quad , \end{aligned} \tag{3.11}$$

whence

$$dH = d(p\dot{q} - L) = \dot{q}dp - \dot{p}dq \ ; \tag{3.12}$$

[1] H. Poincare, *Les Methodes Nouvelles de la Mecanique Celeste*, Gauthier-Villars, Paris (1892).

$$H(p,q,t) = p\dot{q} - L \tag{3.13}$$

is the energy written with p and q; p is obtained from $p = \partial L/\partial \dot{q}$. H is called the hamiltonian of the motion, or the Hamilton's function; from equations (3.12) we get the Hamilton's equations of motion (canonical equations of motion)

$$\dot{q} = \frac{\partial H}{\partial p} \ , \ \dot{p} = -\frac{\partial H}{\partial q} \ ; \tag{3.14}$$

we note that the time is a parameter, which does not enter the above transformation (called Legendre's transformation); q and p are called canonical variables. In addition,

$$\frac{dH}{dt} = \frac{\partial H}{\partial t} + \frac{\partial H}{\partial q}\dot{q} + \frac{\partial H}{\partial p}\dot{p} = \frac{\partial H}{\partial t} \ . \tag{3.15}$$

The motion of any function $f(p,q,t)$ is given by

$$\begin{gathered}\frac{df}{dt} = \frac{\partial f}{\partial t} + \frac{\partial f}{\partial q}\dot{q} + \frac{\partial f}{\partial p}\dot{p} = \\ = \frac{\partial f}{\partial t} + \frac{\partial f}{\partial q}\frac{\partial H}{\partial p} - \frac{\partial f}{\partial p}\frac{\partial H}{\partial q} = \frac{\partial f}{\partial t} + \{H,f\} \ ,\end{gathered} \tag{3.16}$$

where

$$\{H,f\} = \frac{\partial H}{\partial p}\frac{\partial f}{\partial q} - \frac{\partial H}{\partial q}\frac{\partial f}{\partial p} \tag{3.17}$$

is called the Poisson bracket. Similarly,

$$\{f,g\} = \frac{\partial f}{\partial p}\frac{\partial g}{\partial q} - \frac{\partial f}{\partial q}\frac{\partial g}{\partial p} \tag{3.18}$$

is the Poisson bracket of any two functions f and g; since $\{p,q\} = 1$ the canonical variables q and p are called canonical conjugate. Similarly, we have $\{M_i, M_j\} = -\varepsilon_{ijk}M_k$. We note that $\{f,gh\} = \{f,g\}h + g\{f,h\}$, for three functions f, g, h.

The Hamilton's equations are differential equations of the first order, in contrast with Lagrange's equations which are of the second order.

3.3 Hamilton-Jacobi equation

Using the Hamilton's function, the mechanical action can be written as

$$S = \int dtL = \int dt(p\dot{q} - H) = \int (pdq - Hdt) \ , \tag{3.19}$$

where the integration is performed from t_0 to t and from $q_0 = q(t_0)$ to $q = q(t)$; it follows

$$p = \frac{\partial S}{\partial q} \ , \ H = -\frac{\partial S}{\partial t} \ ; \tag{3.20}$$

the action may be viewed as a function of q and t; from equations (3.20) we get

$$\frac{\partial S}{\partial t} + H(\frac{\partial S}{\partial q}, q, t) = 0 \ , \tag{3.21}$$

where p has been replaced by $\partial S/\partial q$ in $H(p, q, t)$. This is the Hamilton-Jacobi equation. The solution of this equation is greatly simplified for conservative motion, where the hamiltonian does not depend on time, $\partial S/\partial t = -E = const$, and separable variables, *i.e.* for canonical couples p_α, q_α appearing in separate functions $f_\alpha(p_\alpha, q_\alpha)$ in H; then, the problem is reduced to the integrals of the motion C_α and the solution is $S(t, q_\alpha, C_\alpha)$.

For a particle

$$E = \frac{1}{2m}p^2 + U(\mathbf{r}) \tag{3.22}$$

and

$$S = -Et + \int pds \ , \tag{3.23}$$

where the integral is taken along the particle's path; the extremum condition

$$\delta S = \delta \int pds = \delta \int \sqrt{2m(E - U(\mathbf{r})}ds = 0 \tag{3.24}$$

determines the path $s(\mathbf{r})$; it is known as the Maupertuis principle; for a free particle it gives the shortest distance. Similarly, we have the velocity $v = ds/dt = \sqrt{2(E - U)/m}$; using v and the path equation, we get the trajectory $\mathbf{r}(t)$.

3.4 Canonical transformations

Let

$$Q = Q(p, q, t) \ , \ P = P(p, q, t) \tag{3.25}$$

be a transformation which preserves the equations of motion, *i.e.*

$$\dot{p} = -\frac{\partial H}{\partial q} \ , \ \dot{q} = \frac{\partial H}{\partial p} \tag{3.26}$$

and

$$\dot{P} = -\frac{\partial H^{'}}{\partial Q} \ , \ \dot{Q} = \frac{\partial H^{'}}{\partial P} \ , \tag{3.27}$$

where $H^{'}$ is the transformed hamiltonian; more precisely, we impose the changes in the mechanical action be the same, *i.e.*

$$\delta \int (pdq - Hdt) = \delta \int (PdQ - H^{'}dt) \ ; \tag{3.28}$$

this condition implies

$$pdq - Hdt = PdQ - H^{'}dt + dF \ , \tag{3.29}$$

or

$$dF = pdq - PdQ + (H^{'} - H)dt \ ; \tag{3.30}$$

hence

$$p = \frac{\partial F}{\partial q} \ , \ P = -\frac{\partial F}{\partial Q} \ , \ H^{'} = H + \frac{\partial F}{\partial t} \ ; \tag{3.31}$$

$F(q, Q, t)$ is called the generating function of the transformation. Similarly, we may write

$$d(F + PQ) = dG = pdq + QdP + (H^{'} - H)dt \tag{3.32}$$

and

$$p = \frac{\partial G}{\partial q} \ , \ Q = \frac{\partial G}{\partial P} \ , \ H^{'} = H + \frac{\partial G}{\partial t} \ , \tag{3.33}$$

with the new generating function $G(q, P, t)$. If F, G do not depend on time, the hamiltonian is preserved. Canonical transformations are used for getting simpler equations of motion.

The motion is described by a canonical transformation. Indeed, let $q = q_t$, $p = p_t$ for some moment of time t. Then $Q = q_{t+\tau} = Q(q,p,t,\tau)$, $P = p_{t+\tau} = P(q,p,t,\tau)$, for some time interval τ. The change of the action is

$$dS = p_{t+\tau}dq_{t+\tau} - p_t dq_t - (H^{'} - H)dt = PdQ - pdq - (H^{'} - H)dt, \quad (3.34)$$

hence the generating function $F = -S$, which satisfies equations (3.31); we can see that $H^{'} = H - \partial S/\partial t$, *i.e.* $-\partial S/\partial t$ is the hamiltonian added during the motion. The action should be viewed here as a function $S(q, q_{t+\tau}, t, \tau)$.

Canonical transformations preserve the Poison brackets. Indeed, g in $\{f,g\}_{p,q}$ may be viewed as a hamiltonian; then $df/dt = \partial f/\partial t - \{f,g\}_{p,q}$, but the time derivatives do not depend on the choice of coordinates; therefore, $\{f,g\}_{p,q} = \{f,g\}_{P,Q}$.

The points (p,q) define a space called phase space; it is a space of mechanical states. Canonical transformations leave unchanged the volume in the space phase. This is called the Liouville theorem. Indeed, the infinitesimal volume $d\Gamma = dpdq$ changes according to the jacobian of the transformation, $d\Gamma = [\partial(p,q)/\partial(P,Q)]dPdQ$, *i.e.* $d\Gamma = [\partial(p,q)/\partial(P,Q)]d\Gamma^{'}$; but $\partial(p,q)/\partial(P,Q) = [\partial(p,q)/\partial(P,q)]/ /[\partial(P,Q)/\partial(P,q)] = [\partial(p)/\partial(P)]/[\partial(Q)/\partial(q)]$; the elements of the first determinant are $\partial p_\alpha/\partial P_\beta = \partial^2 G/\partial q_\alpha \partial P_\beta$ and the elements of the second determinant are $\partial Q_\alpha/\partial q_\beta = \partial^2 G/\partial P_\alpha \partial q_\beta$; the two determinants are equal, so $d\Gamma = d\Gamma^{'}$.

The canonical transformations are helpful in solving the Hamilton-Jacobi equation. Indeed, the solution $S(t, q_\alpha, C_\alpha)$ of the Hamilton-Jacobi equation can be viewed as the generating function of a canonical transformation with the integration constants C_α as momenta P_α; we have $Q_\alpha = \partial S/\partial C_\alpha$ and the new hamiltonian $H^{'} = H + \partial S/\partial t = 0$; therefore, the equations of motion are $\dot{Q}_\alpha = 0$; it follows $\partial S/\partial C_\alpha = Q_\alpha = const$, equations which give the original coordinates q_α as functions of t and C_α and the new constants Q_α ($2s$ integrals of the motion).

3.5 Periodic motion

An interesting case is the conservative periodic motion with one degree of freedom. Let a particle with mass m moving in a potential $U(q)$, where q is the coordinate; for an energy E, we have

$$E = \frac{1}{2m}p^2 + U(q) \tag{3.35}$$

and the motion is restricted to the region defined by the points $q_{1,2}$ given by $E = U(q_{1,2})$; the particle moves forth and back between $q_1$and q_2 with period T and angular frequency $\omega = 2\pi/T$ (in general, depending on E). Leaving aside the temporal contribution, the mechanical action is

$$S(q,E) = \int_{q_0}^{q} pdq' = \int_{q_0}^{q} \sqrt{2m(E - U(q')}dq' \ ; \tag{3.36}$$

it depends on q and E, which offers the possibility to extend the meaning of S. We can see that S is a periodic function with period $\Delta S = 2\pi I$, where

$$I = \frac{1}{2\pi}\oint pdq \ ; \tag{3.37}$$

the integration in equation (3.37) is performed over the entire orbit (trajectory) from q_1 to q_2 and back to q_1. As a function of q, S is a multi-valued function. For reasons given below, the quantity I introduced by equation (3.37) is called adiabatic invariant.[2] It can also be written as

$$I = \frac{1}{2\pi}\iint dpdq \ , \tag{3.38}$$

which is the whole surface Γ covered by the canonical variables in the phase space during the motion. We can also write

$$\frac{\partial I}{\partial E} = \frac{1}{2\pi}\oint \frac{\partial p}{\partial E}dq = \frac{1}{2\pi}\oint \frac{dq}{\dot{q}} = \frac{1}{2\pi}\oint dt = \frac{T}{2\pi} = \frac{1}{\omega} \ , \tag{3.39}$$

[2]P. Ehrenfest, "Adiabatic invariants and the theory of quanta", Phil. Mag. **33** 500 (1917); "On adiabatic changes of a system in connection with the quantum theory", Proc. Acad. Amst. **19** 576 (1917); "Adiabatische Invarianten und Quantentheorie", Ann. Phys. **51** 327 (1916).

or

$$\frac{\partial E}{\partial I} = \omega \ . \tag{3.40}$$

The adiabatic invariant I is constant in time,

$$I = \frac{1}{2\pi}\oint pdq = \frac{1}{2\pi}\oint p\dot{q}dt \ , \tag{3.41}$$

since it is an average over the orbit, as one can see from equation (3.41); it is $2\overline{E}_{kin}/\omega$, where $\overline{E}_{kin}$ is the average of the kinetic energy. Since the motion is periodic with period T, the coordinate and the momentum can be expanded in Fourier series

$$q = \sum_n q_n e^{-in\omega t} \ , \ p = \sum_n p_n e^{-in\omega t} \ , \tag{3.42}$$

where $q_n = q^*_{-n}$, $p_n = -imn\omega q_n$ and n is any integer; we have

$$I = i\sum_n np_n q_{-n} \ ; \tag{3.43}$$

for a harmonic oscillator the Fourier series reduce to $n = \pm 1$ and $I = E/\omega$.

3.6 Angular coordinates

The interesting properties of the adiabatic invariant follow from viewing $S(q,E)$ as a function of $S(q,I)$ which can be taken as the generating function of the canonical transformations $P = I$; we have

$$p = \frac{\partial S}{\partial q} \ , \ \varphi = \frac{\partial S}{\partial I} \ , \tag{3.44}$$

where φ is the variable canonically conjugate to I; it is called the angular coordinate. Using equations (3.36) and (3.40) we get

$$\varphi = \int_{q_0}^{q} \frac{\partial p}{\partial E} dq' \frac{\partial E}{\partial I} = \omega \int_{q_0}^{q} \frac{dq'}{\dot{q}'} = \omega t \ . \tag{3.45}$$

Since S does not depend on time, the transformed hamiltonian $H^{'}$ is the same as the old hamiltonian H and we have the equations of motion

$$\dot{\varphi} = \frac{\partial H}{\partial I} = \frac{\partial E}{\partial I} = \omega \ , \ \dot{I} = \frac{\partial H}{\partial \varphi} = 0 \ ; \tag{3.46}$$

H does not depend on φ and I is constant, as we know; $\varphi = \omega t = 2\pi t/T$ and, during the motion φ varies with multiples of 2π; it is a many-valued function of the mechanical state (defined by p and q); S, p and q, as well as any one-valued function of p and q, can be expanded in a Fourier series in terms of the angular coordinate φ, *e.g.*

$$q = \sum_n q_n e^{-in\varphi} \ ; \tag{3.47}$$

if the original coordinate q is itself an angular coordinate, any trigonometric function of it is one-valued.

The angular coordinate and the adiabatic invariant do not help much to solve the equations of motion, but they offer interesting insights into the nature of the motion.

Indeed, if we consider a finite motion with several degrees of freedom and the variables are separable, then, according to the Hamilton-Jacobi equation, the action can be written as a sum

$$S = \sum_\alpha S_\alpha(q_\alpha) \ , \tag{3.48}$$

where we omit the constants related to the energy E. We may define p_α and adiabatic invariants I_α for each degree of freedom; the motion is periodic for each coordinate, providing the remaining coordinates are "frozen". Therefore, we can introduce angular variables φ_α and, due to the periodicity, we may write for any one-valued function of q and p a multiple Fourier series

$$F = \sum_{n_1, n_2, ...n_s} F(n_1, n_2, ...n_s) e^{-i(n_1\varphi_1 + n_2\varphi_2 + ...n_s\varphi_s)} \ , \tag{3.49}$$

or

$$F = \sum_{n_1, n_2, ...n_s} F(n_1, n_2, ...n_s) e^{-i(n_1\omega_1 + n_2\omega_2 + ...n_s\omega_s)t} \ , \tag{3.50}$$

where s is the number of degrees of freedom. Since, in general, the frequencies ω_α are incommensurate, the above function is not periodic, either as a whole, or in any coordinate.[3] We may imagine that after a sufficiently long time, for many particles, the orbits may pass sufficiently close to any point, which is the so-called ergodic theorem. If the frequencies are commensurate, all or some, for some I_α, the motion is called degenerate. A degeneracy may appear for instance for $l_1\omega_1 - l_2\omega_2 = 0$, where $l_{1,2}$ are integers; any trigonometric function of this quantity may be viewed as a one-valued integral of the motion; $I_{1,2}$ appear in energy only in combination $l_2 I_1 + l_1 I_2$; I_α may be viewed as s (one-valued) integrals of the motion; therefore, for a degenerate motion there exist additional useful integrals of the motion (combinations of the form $\varphi_\alpha\omega_\beta - \varphi_\beta\omega_\alpha$ are many-valued).

3.7 Adiabatic invariants

Let us assume that the conservative hamiltonian of a periodic motion acquires a slow time perturbation, such that its variation in a period is much smaller than its value. Then, energy acquires a slight time dependence, $E \to E(t)$ and momentum can be written as a function $p \to p(E(t), t)$. The typical example is a harmonic oscillator with the frequency varying slowly in time. The time derivative of the adiabatic invariant is

$$\frac{dI}{dt} = \frac{1}{2\pi}\oint\left(\frac{\partial p}{\partial E}\frac{\partial E}{\partial t} + \frac{\partial p}{\partial t}\right)dq \ ; \tag{3.51}$$

making use of the slow variation of the perturbation and averaging over the periodic motion, we get

$$\overline{\frac{dI}{dt}} = \frac{1}{2\pi}\oint\left(\frac{\partial E}{\partial q} - \frac{\partial E}{\partial q}\right)dq = 0 \ ; \tag{3.52}$$

we can see that the averaged adiabatic invariant is constant for a slow perturbation; slow perturbations are called adiabatic. The average is necessary in equation (3.52) since we preserve the orbit.

[3] Only Newton's gravitation law and the space harmonic oscillator exhibit closed periodic orbits.

A typical slow time-variation implies for any quantity f an equation of the form

$$\frac{df}{dt} = f_1 - \frac{f}{\tau} = f_1 - \gamma f \tag{3.53}$$

for $t > 0$, where $\tau = 1/\gamma \gg t$, and

$$\frac{df}{dt} = f_2 + \gamma f \tag{3.54}$$

for $t < 0$. The solution is

$$f = \begin{cases} f_\infty - \frac{1}{2}(f_\infty - f_{-\infty})e^{-\gamma t} \ , \\ f_{-\infty} + \frac{1}{2}(f_\infty - f_{-\infty})e^{\gamma t} \ , \end{cases} \tag{3.55}$$

where $f_{\pm\infty} = f(t \to \pm\infty)$. The occurrence of a time-decaying exponential $e^{\mp\gamma t}$ means an imaginary part of the frequency $\Omega = n\omega$ in the Fourier series, $\Omega \to \Omega - i\gamma$ for $t > 0$ (lower half-plane of the complex frequency) and $\Omega \to \Omega + i\gamma$ for $t < 0$ (upper half-plane of the complex frequency). This imaginary part means a corresponding pole in the Fourier transform.

If the mechanical action acquires a slow dependence on the time, then the hamiltonian for the angular variables is

$$H^{'} = H + \frac{\partial S}{\partial t} = H - h \ , \tag{3.56}$$

where we denoted by h the additional term brought to the hamiltonian by the time dependence of the action; the canonical equations of motion are

$$\dot{\varphi} = \omega - \frac{\partial h}{\partial I} \ , \ \dot{I} = \frac{\partial h}{\partial \varphi} \tag{3.57}$$

where $\varphi = \partial S / \partial I$. We can see that h is a (slowly-varying) periodic function of φ, which, averaged over the orbit, gives zero, *i.e.* $\overline{\dot{I}} = 0$, as expected. The change in the adiabatic invariant is

$$\Delta I = \int_{-\infty}^{+\infty} \frac{\partial h}{\partial \varphi} dt \tag{3.58}$$

and, since $\partial h/\partial\varphi$ is a periodic function of φ, we can write

$$\Delta I = -i\sum_n nh_n \int \frac{e^{-in\varphi}}{\omega}d\varphi \ ; \tag{3.59}$$

the pole φ_0 given by $\omega(\varphi_0) = 0$ gives the decay law $\Delta I \sim^{-nIm\varphi_0}$. Since $\omega = d\varphi/dt$, it follows

$$\varphi_0 = \int^{t_0} \omega dt \ , \tag{3.60}$$

where the complex time t_0 is given by $\varphi_0 = \omega t_0$; the lower limit in the integral in equation above is any real t; we have therefore

$$\Delta I \sim e^{-Im\int^{t_0}\omega dt} \ , \tag{3.61}$$

where we take the slowest time dependence; the adiabatic variation means large $Im\varphi_0$.

3.8 Harmonic oscillator

The harmonic oscillator with one degree of freedom is defined by the hamiltonian

$$H = \frac{1}{2m}p^2 + \frac{1}{2}m\omega^2 q^2 \ , \tag{3.62}$$

where m is the mass and ω is the frequency; the solution of the equations of motion are

$$q = A\cos(\omega t + \alpha) \ , \ p = -mA\omega\sin(\omega t + \alpha) \ , \tag{3.63}$$

where the amplitude A and the phase α are determined from the initial conditions. The energy is $E = mA^2\omega^2/2$ and the adiabatic invariant is $I = E/\omega$; therefore,

$$q = \sqrt{\frac{2I}{m\omega}}\cos(\omega t + \alpha) \ , \ p = -\sqrt{2m\omega I}\sin(\omega t + \alpha) \tag{3.64}$$

the action is

$$S = \frac{1}{2}I(\varphi - \frac{1}{2}\sin 2\varphi) \ , \tag{3.65}$$

where $\varphi = \omega t$ is the angular variable.

In three dimensions

$$H = \frac{1}{2m}p_x^2 + \frac{1}{2m}p_y^2 + \frac{1}{2m}p_z^2 + \frac{1}{2}m\omega_x^2 x^2 + \frac{1}{2}m\omega_y^2 y^2 + \frac{1}{2}m\omega_z^2 z^2 \; ; \qquad (3.66)$$

the separation of variables leads to three independent one-dimensional oscillators with energies $E_{x,y,z}$, $E = E_x + E_y + E_z$. The frequencies $\omega_{x,y,z}$ are not commensurate in general and the trajectory is not closed; the degenerate case $\omega_x = \omega_y = \omega_z$ is called the space oscillator; its motion is closed.

3.9 Kepler's problem

3.9.1 Spherical coordinates

With usual notations the velocity of a particle in spherical coordinates is

$$\mathbf{v} = (\dot{r},\, r\dot{\theta},\, r\sin\theta\dot{\varphi}) \; ; \qquad (3.67)$$

the lagrangian of a free particle with mass m reads

$$L = \frac{1}{2}m(\dot{r}^2 + r^2\dot{\theta}^2 + r^2\sin^2\theta\dot{\varphi}^2) \; ; \qquad (3.68)$$

the momenta are

$$p_r = m\dot{r} \; , \; p_\theta = mr^2\dot{\theta} \; , \; p_\varphi = mr^2\sin^2\theta\dot{\varphi} \qquad (3.69)$$

and the hamiltonian is

$$H = \frac{1}{2m}\left(p_r^2 + \frac{p_\theta^2}{r^2} + \frac{p_\varphi^2}{r^2\sin^2\theta}\right) \; . \qquad (3.70)$$

If the potential U is of the form

$$U = u_1(r) + \frac{u_2(\theta)}{r^2} + \frac{u_3(\varphi)}{r^2\sin^2\theta} \qquad (3.71)$$

the Hamilton-Jacobi equation is separable; indeed, we have

$$\begin{gathered} p_\varphi^2 + 2mu_3(\varphi) = \alpha \ , \ p_\theta^2 + \tfrac{\alpha}{\sin^2\theta} + 2mu_2(\theta) = \beta \ , \\ p_r^2 + \tfrac{\beta}{r^2} + 2mu_1(r) = 2mE \ , \end{gathered} \tag{3.72}$$

where α, β are constants and E is the energy; they are integrals of the motion.

3.9.2 Centrally symmetric field

The problem is simplified for a central field, when U reduces to $U(r)$. Then, the angular momentum $\mathbf{M} = \mathbf{r} \times \mathbf{p}$ is a constant vector, $\dot{\mathbf{M}} = 0$. Consequently, the motion proceeds in a plane perpendicular to $\mathbf{M}$, where we use polar coordinates r and φ; the velocities are $\dot{r}$ and $r\dot{\varphi}$ and the angular momentum $\mathbf{M}$ reduces to its component perpendicular to the plane which contains the trajectory, $M = mr^2\dot{\varphi}$. The lagrangian is

$$L = \frac{1}{2}m(\dot{r}^2 + r^2\dot{\varphi}^2) - U(r) \ , \tag{3.73}$$

the momenta are

$$p_r = m\dot{r} \ , \ p_\varphi = mr^2\dot{\varphi} = M \tag{3.74}$$

and the hamiltonian is

$$H = \frac{1}{2m}p_r^2 + \frac{M^2}{2mr^2} + U \ . \tag{3.75}$$

The equations of motion

$$\tfrac{d}{dt}(mr^2\dot{\varphi}) = 0 \ , \ \tfrac{d}{dt}(m\dot{r}) = -\tfrac{\partial U}{\partial r} + \tfrac{M^2}{mr^3} \tag{3.76}$$

lead to the energy conservation

$$E = \frac{1}{2m}p_r^2 + \frac{M^2}{2mr^2} + U \ , \tag{3.77}$$

or

$$\dot{r} = \sqrt{\frac{2}{m}(E - U) - M^2/m^2r^2} \ ; \tag{3.78}$$

making use of $\dot{\varphi} = M/mr^2$ we get

$$\frac{d\varphi}{dr} = \frac{1/r^2}{\sqrt{\frac{2m}{M^2}(E-U) - 1/r^2}} \ , \tag{3.79}$$

which gives the parametric equation of the trajectory $\varphi(r)$.

The above equations show that φ is a monotonically increasing function of the time ($\dot{\varphi} = M/mr^2 > 0$); the coordinate r moves in the effective potential

$$U_{eff} = \frac{M^2}{2mr^2} + U \ , \tag{3.80}$$

which includes the centrifugal contribution $M^2/2mr^2$. The motion along the coordinate r is limited by the condition $E = U_{eff}(r)$; if there exists one solution r_{min} of this equation, the motion is infinite, $r > r_{min}$; if there exist two solutions r_{min} and r_{max} of this equation, the motion is finite, $r_{min} < r < r_{max}$. The points r_{min} and r_{max} are turning points, where $\dot{r} = 0$ (but, of course, $\dot{\varphi} \neq 0$). The change $\Delta\varphi$ in the angle for a motion from r_{min} forth to r_{max} and back to r_{min} is not a rational fraction of 2π, in general; therefore, trajectory is not closed. For $U \sim 1/r$ (Kepler's problem) and $U \sim r^2$ (space oscillator) the orbits are closed. Since

$$\frac{1}{2}m\dot{r}^2 = E - U - \frac{M^2}{2mr^2} > 0 \ , \tag{3.81}$$

it follows that the particle may fall to the centre for

$$(r^2U)_{r\to 0} < -M^2/2m \ , \tag{3.82}$$

i.e. either $U \sim -\alpha/r^2$ with $\alpha > M^2/2m$, or $U \sim -1/r^\beta$ with $\beta > 2$.

3.9.3 The potential $U = -\alpha/r$

Let us assume $U = -\alpha/r$, where, α is a constant; first, we assume $\alpha > 0$. This is Kepler's problem, for the gravitational attractive law or the Coulomb attractive law of two particles with opposite electric charges. From

$$E = \frac{1}{2}m\dot{r}^2 + U_{eff} = \frac{1}{2}m\dot{r}^2 + \frac{M^2}{2mr^2} - \frac{\alpha}{r} \tag{3.83}$$

we can see that the minimum of the U_{eff} is $-m\alpha^2/2M^2$, which shows that the energy should be greater than this value, $E > -m\alpha^2/2M^2$. For $-m\alpha^2/2M^2 < E < 0$ the motion is finite, covering the region between $r_{min} = (\alpha/2 \mid E \mid)[1 - \sqrt{1 - 2 \mid E \mid M^2/m\alpha^2}]$ and $r_{max} = (\alpha/2 \mid E \mid)[1 + \sqrt{1 - 2 \mid E \mid M^2/m\alpha^2}]$; for $E > 0$ the motion is infinite. Equation (3.79) can be integrated immediately by the change of variable $u = 1/r - m\alpha/M^2$; we get

$$p/r = 1 + e\cos\varphi \ , \tag{3.84}$$

where

$$p = M^2/m\alpha \ , \ e = \sqrt{1 + 2EM^2/m\alpha^2} \ . \tag{3.85}$$

This is the parametric equation of a conics with the origin at one focus and eccentricity e. For $E < 0$ the curve is an ellipse with $\varphi = 0$ the perihelion (r_{min}) and $\varphi = \pi$ the aphelion (r_{max}) (this is a Kepler's law); in cartesian coordinates the equation can be written as

$$\frac{(x+f)^2}{a^2} + \frac{y^2}{b^2} = 1 \ , \tag{3.86}$$

where $a = p/(1 - e^2) = \alpha/2 \mid E \mid$ is the major semi-axis, $b = p/\sqrt{1 - e^2} = M/\sqrt{2m \mid E \mid}$ is the minor semi-axis and $f = ae$ is the focal distance; the perihelion is $r_{min} = a(1 - e)$ and the aphelion is $r_{max} = a(1 + e)$. The equation $M = mr^2\dot{\varphi}$ shows that $M = 2m\dot{A}$, where A is the area swept by the radius r (this is another Kepler's law) ; for an ellipse $A = \pi ab$, so that $MT = 2\pi mab$, or

$$T^2 = \frac{4\pi^2 m}{\alpha} a^3 = \frac{\pi^2 m\alpha^2}{2 \mid E \mid^3} \ , \tag{3.87}$$

where T is the period; this is another Kepler's law (the period squared is proportional to the cube of the major semi-axis). From equation (3.78) we get

$$\begin{gathered} r = a(1 - e\cos\xi) \ , \ t = \sqrt{ma^3/\alpha}(\xi - e\sin\xi) \ , \\ x = a(\cos\xi - e) \ , \ y = a\sqrt{1 - e^2}\sin\xi \ . \end{gathered} \tag{3.88}$$

Making use of r_{min} and r_{max}, we get $r_{min} + r_{max} = \alpha / \mid E \mid = 2a$ and $r_{min} r_{max} = M^2/2m \mid E \mid = a^2(1-e^2)$; hence,

$$E = \frac{\alpha}{2a} \ , \ E = -\frac{m\alpha^2(1-e^2)}{2M^2} \ . \tag{3.89}$$

For $E > 0$ the motion is infinite and the orbit is a hyperbola; $r_{min} = a(e-1)$; the trajectory is given by

$$\begin{gathered} r = a(\cosh\xi - 1) \ , \ t = \sqrt{ma^3/\alpha}(e\sinh\xi - \xi) \ , \\ x = a(e - \cosh\xi) \ , \ y = a\sqrt{e^2-1}\sinh\xi \ . \end{gathered} \tag{3.90}$$

For $E = 0$, $e = 1$ and the orbit is a parabolla, with $r_{min} = p/2$. The trajectory is given by

$$\begin{gathered} r = \frac{1}{2}p(1+\xi^2) \ , \ t = \sqrt{mp^3/\alpha} \cdot \frac{1}{2}\xi(1 + \frac{1}{3}\xi^2 0 \ , \\ x = \frac{1}{2}p(1-\xi^2) \ , \ y = p\xi \ . \end{gathered} \tag{3.91}$$

For $\alpha < 0$ the potential is repulsive and the motion is possible $E > 0$; the orbit is a hyperbola

$$p/r = -1 + e\cos\varphi \tag{3.92}$$

with $r_{min} = a(e+1)$ and the trajectory

$$\begin{gathered} r = a(\cosh\xi + 1) \ , \ t = \sqrt{ma^3/\alpha}(e\sinh\xi + \xi) \ , \\ x = a(e + \cosh\xi) \ , \ y = a\sqrt{e^2-1}\sinh\xi \ . \end{gathered} \tag{3.93}$$

3.9.4 Adiabatic invariants for central field

For a central field the mechanical action is

$$S = -Et + p_\varphi \varphi + \int^r drp_r \ , \tag{3.94}$$

where $p_\varphi = M = const$; it follows that the adiabatic invariant for the coordinate φ is

$$I_\varphi = \frac{1}{2\pi}\int_0^{2\pi} d\varphi' M = M \tag{3.95}$$

and the adiabatic invariant for the coordinate r

$$I_r = \frac{1}{2\pi}2\int_{r_{min}}^{r_{max}} drp_r = \\ = \frac{1}{\pi}\int_{r_{min}}^{r_{max}} dr\sqrt{2m(-\mid E\mid -U)-M^2/r^2} \ , \tag{3.96}$$

according to equation (3.72). Making use of $p_r = m\dot{r}$ and equation (3.84), we have

$$I_r = \frac{m}{\pi}\int_{\pi}^{0} d\varphi r'^2\dot{\varphi} = \frac{M}{\pi}\int_{\pi}^{0} d\varphi\frac{r'^2}{r^2} = \\ = \frac{Me^2}{\pi}\int_{\pi}^{0} d\varphi\frac{\sin^2\varphi}{(1+e\cos\varphi)^2} \ ; \tag{3.97}$$

integrating by parts, we get

$$I_r = -M + \frac{M}{\pi}\int_0^{\pi/2} d\varphi\left(\frac{1}{1+e\cos\varphi}+\frac{1}{1-e\cos\varphi}\right) \ ; \tag{3.98}$$

the integrals in this equation can be carried out by the substitution $\tan\varphi/2 = t$; finally, we get

$$I_r = -M + \frac{M}{\sqrt{1-e^2}} = -M + \sqrt{\frac{m\alpha^2}{2\mid E\mid}} \ . \tag{3.99}$$

It follows

$$1 - e^2 = \frac{I_\varphi^2}{(I_r + I_\varphi)^2} \tag{3.100}$$

and

$$E = -\frac{m\alpha^2}{2(I_r + I_\varphi)^2} \ . \tag{3.101}$$

The parameter p is given by $p = I_\varphi^2/m\alpha$. When m, α vary slowly, the eccentriciy of the orbit remains unchanged, while its dimensions vary according to the parameter p.

3.9.5 Lenz vector

Let us consider the potential $U = -\alpha/r$, where α is either positive, or negative. Then, the Lenz vector[4]

$$\mathbf{A} = \mathbf{v}\times\mathbf{M} - \frac{\alpha\mathbf{r}}{r} \tag{3.102}$$

[4]Known also as the Laplace-Runge-Lenz vector.

is a constant of motion; indeed, using

$$\frac{d\mathbf{A}}{dt} = \dot{\mathbf{v}} \times (\mathbf{r} \times m\mathbf{v}) - \frac{\alpha\mathbf{v}}{r} + \frac{\alpha(\mathbf{rv})}{r^3} \tag{3.103}$$

and the equation of motion $m\dot{\mathbf{v}} = -\alpha\mathbf{r}/r^3$, we get$\dot{\mathbf{A}} = 0$. The occurrence of an additional one-valued integral of the motion $\mathbf{A}$, beside E and $\mathbf{M}$, is a consequence of the degeneracy of the potential $\sim 1/r$. Making use of equations (3.88)-(3.93) we find easily that the vector $\mathbf{A}$ lies in the plane of the motion, is directed from the focus to the perihelion (along the x-axis) and has the magnitude αe; in addition, $\mathbf{AM} = 0$, $\mathbf{rA} = \alpha er\cos\varphi$, or

$$\mathbf{rA} = \frac{1}{m}M^2 - \alpha r \tag{3.104}$$

and, making use of equation (3.89),

$$A^2 = \frac{2}{m}M^2E + \alpha^2 \ . \tag{3.105}$$

From equation (3.102) we have also

$$A = Mv_y - \frac{\alpha x}{r} \ , \ -Mv_x - \frac{\alpha y}{r} = 0 \ , \tag{3.106}$$

hence

$$p_x^2 + (p_y - A/M)^2 = m\alpha^2/M^2 \ , \tag{3.107}$$

where $p_{x,y}$ are the cartesian momenta. The existence of an additional integral of the motion helps solve the equations of motion: from equation (3.104) we get immediately the equation of the trajectory (equation (3.84)).

3.10 Electric and magnetic fields

3.10.1 Centre of mass

Let $\mathbf{r}_a$ be the position of the a-th particle with mass m_a; the lagrangian of such an assembly of particles can be of the form

$$L = \sum_a \frac{1}{2}m_a v_a^2 - \frac{1}{2}\sum_{a\neq b} U(\mathbf{r}_a - \mathbf{r}_b) - \sum_a V(\mathbf{r}_a) \ , \tag{3.108}$$

where $v_a = \dot{\mathbf{r}}_a$ is the velocity, U is the interaction and V is an external interaction which depends only on the coordinates. We introduce the relative coordinates $\mathbf{u}_a$ through

$$\mathbf{r}_a = \mathbf{r} + \mathbf{u}_a \quad , \tag{3.109}$$

where $\mathbf{r}$ is the position of the centre of mass and $\sum_a m_a \mathbf{u}_a = 0$; hence,

$$\mathbf{r} = \frac{1}{M}\sum_a m_a \mathbf{r}_a \quad , \quad \mathbf{u}_a = \mathbf{r}_a - \frac{1}{M}\sum_a m_a \mathbf{r}_a \quad , \tag{3.110}$$

where $M = \sum_a m_a$ is the mass of the assembly; the lagrangian becomes

$$L = \frac{1}{2}M\dot{\mathbf{r}}^2 + \sum_a \frac{1}{2}m_a \dot{\mathbf{u}}_a^2 - \frac{1}{2}\sum_{a,b}' U(\mathbf{u}_a - \mathbf{u}_b) - \sum_a V(\mathbf{r} + \mathbf{u}_a) \; . \tag{3.111}$$

We can see that the centre of mass is decoupled from the relative motion as long as the external field is absent. If $u_a \ll r$, we may write

$$V(\mathbf{r} + \mathbf{u}_a) = V(\mathbf{r}) + \mathbf{u}_a gradV(\mathbf{r}) + \frac{1}{2}u_{ai}u_{aj}\partial_i\partial_j V(\mathbf{r}) + ... \quad , \tag{3.112}$$

where i, j denote the cartesian coordinates. We may solve for the motion of the center of mass by neglecting the **u**-coupling, and use this solution in the **u**-coupling for the **u**-motion as an external time-dependent interaction. The first-derivatives term in equation (3.112) is called the dipole interaction and the second-derivatives term is called the quadrupole interaction.

3.10.2 Electromagnetic field

The interaction of the electric charges q_a with an electromagnetic field with scalar potential $\Phi(\mathbf{r}, t)$ and vector potential $\mathbf{A}(\mathbf{r}, t)$, $(1/c)\partial\Phi/\partial t + div\mathbf{A} = 0$, is

$$L_{int} = \sum_a \left[-q_a \Phi(\mathbf{r}_a, t) + q_a \frac{\mathbf{v}_a}{c}\mathbf{A}(\mathbf{r}_a, t) \right] \quad , \tag{3.113}$$

where c is the speed of light in vacuum; the kinetic part of non-relativistic particles with mass m_a is

$$L_{kin} = \sum_a \frac{1}{2} m_a v_a^2 \ ; \tag{3.114}$$

hence we get the electromagnetic momentum

$$\mathbf{p}_a = m\mathbf{v}_a + \frac{q_a}{c}\mathbf{A}(\mathbf{r}_a, t) \tag{3.115}$$

and the standard non-relativistic hamiltonian

$$H = \sum_a \frac{1}{2m_a}\left[\mathbf{p}_a - \frac{q_a}{c}\mathbf{A}(\mathbf{r}_a, t)\right]^2 + \sum_a q_a \Phi(\mathbf{r}_a, t) \ ; \tag{3.116}$$

we note that $\mathbf{p}_a$ includes the electromagnetic momentum $q_a\mathbf{A}(\mathbf{r}_a, t)/c$ beside the mechanical momentum $m\mathbf{v}_a$.
The equations of motion are

$$\begin{aligned} \tfrac{d}{dt}[m_a\mathbf{v}_a + \tfrac{q_a}{c}\mathbf{A}(\mathbf{r}_a, t)] &= -q_a grad\Phi(\mathbf{r}_a, t) + \\ &+ \tfrac{q_a}{c} grad[\mathbf{v}_a\mathbf{A}(\mathbf{r}_a, t)] \ , \end{aligned} \tag{3.117}$$

or

$$\begin{aligned} \tfrac{d}{dt} m_a\mathbf{v}_a &= -q_a grad\Phi(\mathbf{r}_a, t) - \\ &- \tfrac{q_a}{c}\tfrac{\partial \mathbf{A}(\mathbf{r}_a, t)}{\partial t} + \tfrac{q_a}{c}[\mathbf{v}_a \times curl\mathbf{A}(\mathbf{r}_a, t)] \ , \end{aligned} \tag{3.118}$$

where we use here

$$grad(\mathbf{v}\mathbf{A}) = (\mathbf{v}grad)\mathbf{A} + \mathbf{v} \times curl\mathbf{A} \ ; \tag{3.119}$$

in equation (3.118) we recognize the electric field

$$\mathbf{E} = -grad\Phi - \frac{1}{c}\frac{\partial \mathbf{A}}{\partial t} \ , \tag{3.120}$$

the magnetic field

$$\mathbf{H} = curl\mathbf{A} \tag{3.121}$$

and the Lorentz force

$$\mathbf{F} = q\mathbf{E} + \frac{q}{c}\mathbf{v} \times \mathbf{H} \ . \tag{3.122}$$

The electromagnetic field introduced above can be produced both by external charges and charges q_a themselves, whose motion we are interested in. We may compute the electromagnetic field of the charges q_a in terms of their positions and velocities and use it in the lagrangian given above; doing so, it turns out that we cannot get a lagrangian in terms of positions and velocities only, except up to second-order terms v^2/c^2 (without first-order terms); in higher orders of v/c the self-field of the charges cannot be eliminated. Therefore, the expression of the lagrangian only with positions and velocities of the particles is due to the infinite velocity of propagation of the interaction; because of the finite velocity of the electromagnetic field, the field cannot be completely eliminated from the lagrangian.[5] To the lowest-order in v/c the lagrangian includes only the Coulomb interaction,

$$L_{int} = \sum_a \frac{1}{2} m_a v_a^2 - \frac{1}{2} \sum_{a \neq b} \frac{q_a q_b}{|\mathbf{r}_a - \mathbf{r}_b|} . \tag{3.123}$$

3.10.3 Constant electric field

A constant electric field $\mathbf{E}(\mathbf{r}_a)$ is derived from a scalar potential $\mathbf{E} = -grad\Phi$, where Φ is a Coulomb potential. Making use of the relative coordinates and the centre of mass, we get the lagrangian of the relative motion

$$\begin{gathered} L = \frac{1}{2} \sum_a m_a \dot{u}_a^2 - \sum_a q_a \Phi(\mathbf{r}) - \sum_a q_a \mathbf{u}_a grad\Phi(\mathbf{r}) - \\ - \frac{1}{6} \sum_a q_a (3u_{ai}u_{aj} - \delta_{ij}u_a^2)\partial_i\partial_j\Phi(\mathbf{r}) + ... ; \end{gathered} \tag{3.124}$$

the δ_{ij}-term is added for convenience, since $\Delta\Phi = 0$. $d_a = q_a \mathbf{u}_a$ is the vector of the dipole moment of the a-th charge and the tensor

$$Q_{aij} = 3u_{ai}u_{aj} - \delta_{ij}u_a^2 \tag{3.125}$$

is the quadrupole moment of the a-th charge. We may write also

$$\begin{gathered} L = \frac{1}{2} \sum_a m_a \dot{u}_a^2 - \sum_a q_a \Phi(\mathbf{r}) + \sum_a q_a \mathbf{u}_a \mathbf{E}(\mathbf{r}) + \\ + \frac{1}{6} \sum_a q_a (3u_{ai}u_{aj} - \delta_{ij}u_a^2)\partial_i E_j(\mathbf{r}) + ... \end{gathered} \tag{3.126}$$

[5] C. G. Darwin, "The dynamical motions of charged particles", Phil. Mag., Series 6, **39** 537 (1920); L. Landau and E. Lifshitz, *Course of Theoretical Physics*, vol. 2, *The Classical Theory of Fields*, Elsevier, Oxford (1975).

It is easy to see that the total dipole is independent of $\mathbf{r}$ if the total charge is zero; and the quadrupole is independent of $\mathbf{r}$ if the total charge and the total dipole are zero. The dipole energy is

$$V_1 = -\mathbf{dE}(\mathbf{r}) \tag{3.127}$$

and the force acting upon the dipole is

$$\mathbf{F}_1 = \sum_a q_a \mathbf{E} + (\mathbf{d}grad)\mathbf{E} \ ; , \tag{3.128}$$

the first term is vanishing for zero total charge, while the second term relates to the coordinates $\mathbf{r}$. The torque is

$$\mathbf{K} = \sum_a q_a \mathbf{u}_a \times \mathbf{E} = \mathbf{d} \times \mathbf{E} \ , \tag{3.129}$$

for $\sum_a q_a = 0$.

The electric field produced by the charges q_a at $\mathbf{R}$, where $R \gg u_a$, is

$$\mathbf{E}_d = -grad \sum_a \frac{\mathbf{d}_a \mathbf{R}_0}{R_0^3} = \frac{3(\mathbf{dR}_0)\mathbf{R}_0 - \mathbf{d}R_0^2}{R_0^5} \ , \tag{3.130}$$

for $\sum_a q_a = 0$, where $\mathbf{R}_0 = \mathbf{R} - \mathbf{r}$. The energy of another dipole $\mathbf{d}'$ in this field is

$$U_d = -\mathbf{d}'\mathbf{E}_d = \frac{\mathbf{dd}' R_0^2 - 3(\mathbf{dR}_0)(\mathbf{d}'\mathbf{R}_0)}{R_0^5} \ ; \tag{3.131}$$

this is dipole-dipole interaction.

3.10.4 Electric moments

Many atomic particles are composite particles, consisting of electric charges. Atoms and atomic nuclei have not a dipole moment in their ground states, due to the invariance to spatial inversion; however, they may have an induced dipole, caused by an external electric field. Usually, molecules have a dipole moment. The quadrupole moment is more frequent, *e.g.* some nuclei do have a quadrupole moment, though small. An external electric field can induce a dipole moment, and the charges are then polarized electrically, they have a polarizability; in condensed matter the dipole moment of the unit volume is the electric polarization. The electron has not a dipole. The dimension of the atoms is of the order of the Bohr radius $a_H = \hbar^2/me^2 = 0.5\text{Å}$, where $\hbar = 10^{-27} erg \cdot s$ is the Planck's constant, $m = 10^{-27} g$ is the electron mass and $e = -4.8 \times 10^{-10} esu$ (*statcoulomb*) is the electron charge; the atomic dipole is of the order $-ea_H = 2.4 \times 10^{-18} esu$. The dimension of the atomic nuclei is of the order 10^{-13}Å. The dipole-dipole interaction, or dipole or quadrupole interactions with charges give rise to interatomic or intermolecular forces. The interaction of the electrons with the atomic nucleus is called the hyperfine interaction. (The spin-orbit and spin-spin interaction are fine interactions).

The lagrangian of a dipole $\mathbf{d}$ with mass m in an external electric field $\mathbf{E}$ is

$$L = \frac{1}{2} m \dot{u}^2 + \mathbf{dE} \quad , \tag{3.132}$$

where $\mathbf{d} = q\mathbf{u}$, q being the electric charge; $\dot{\mathbf{u}}$ is the velocity $\mathbf{v}$; the hamiltonian is

$$H = \frac{1}{2m} p^2 - \mathbf{dE} \quad , \tag{3.133}$$

where momentum $\mathbf{p}$ is $\mathbf{p} = m\dot{\mathbf{u}}$. The motion of the position $\mathbf{u}$ may imply vibrations and rotations (apart from the motion of the centre-of-mass of the particle). The vibrations require a restoring internal elastic force. The moments are distributed thermally in condensed matter, and the external electric field polarizes them; this is the Curie-Langevin-Debye orientational polarizability.[6]

A dipole may be viewed as a spherical pendulum (spatial, rigid rotator, spherical top), consisting of a point of mass m and charge q which rotates freely in space at the end of a rigid radius $\mathbf{u} = u(\sin\theta\cos\varphi, \sin\theta\sin\varphi, \cos\theta)$, in spherical coordinates; for an electric

[6] P. Curie, "Lois experimentales du magnetisme. Proprietes magnetiques des corps a diverses temperatures", Ann. Chim. Phys. **5** 289 (1895); P. Langevin, "Sur la theorie du magnetisme", J. Physique **4** 678 (1905); "Magnetism et theorie des electrons", Ann. Chim. Phys. **5** 70 (1905); P. Debye, "Einige Resultate einer kinetischen Theorie der Isolatoren", Phys. Z. **13** 97 (1912).

field $\mathbf{E} = E(0,0,1)$ the hamiltonian is

$$L = \frac{1}{2}m\dot{\mathbf{u}}^2 + quE\cos\theta = \frac{1}{2}mu^2(\dot{\theta}^2 + \dot{\varphi}^2\sin^2\theta) + dE\cos\theta \ ; \quad (3.134)$$

$I = mu^2$ is the moment of inertia. The equations of motion are

$$\frac{d}{dt}(I\dot{\varphi}\sin^2\theta) = 0 \ , \ \ I\ddot{\theta} = I\dot{\varphi}^2\sin\theta\cos\theta - dE\sin\theta \ . \quad (3.135)$$

The angular momentum $\mathbf{M} = \mathbf{u} \times m\dot{\mathbf{u}}$ has the components $L_x = Ml^2(-\dot{\theta}\sin\varphi - \dot{\varphi}\sin\theta\cos\theta\cos\varphi)$, $L_y = Ml^2(\dot{\theta}\cos\varphi - \dot{\varphi}\sin\theta\cos\theta\sin\varphi)$, $L_z = Ml^2\dot{\varphi}\sin^2\theta$

$$M_x = I(-\dot{\theta}\sin\varphi - \dot{\varphi}\sin\theta\cos\theta\cos\varphi) \ ,$$

$$M_y = I(\dot{\theta}\cos\varphi - \dot{\varphi}\sin\theta\cos\theta\sin\varphi) \ , \quad (3.136)$$

$$M_z = I\dot{\varphi}\sin^2\theta \ ,$$

and the lagrangian given by equation (3.134) can be written as

$$L = \frac{1}{2}I(\dot{\theta}^2 + \dot{\varphi}^2\sin^2\theta) + dE\cos\theta = \frac{1}{2I}M^2 + dE\cos\theta \ . \quad (3.137)$$

We can see that $M_z = I\dot{\varphi}\sin^2\theta$ is conserved and the lagrangian becomes

$$L = \frac{1}{2}I\dot{\theta}^2 + \frac{M_z^2}{2I\sin^2\theta} + dE\cos\theta \ ; \quad (3.138)$$

the equations of motion are the equations of motion for the angular momentum provided by the torque of the force. For weak field the motion is about the minimum of the effective potential $V_{eff} = M_z^2/2I\sin^2\theta$, which is located at $\theta = \theta/2$; the motion is a small oscillation of the zenithal angle θ about the equatorial plane with frequency $\Omega = M_z/I$ (nutation) and an equatorial rotation (precession) of the azimuthal angle $\varphi = \Omega t$; since the frequency is the same we can rotate the equatorial plane such that the motion is a rotation in a plane. Within this approximation the total angular momentum is conserved. The energy M_z^2/I is of the order of the temperature T; it follows $\Omega \simeq \sqrt{T/I}$; for $T = 300K = 4 \times 10^{-14}erg$ $(1K = 1.38 \times 10^{-16}erg)$

and $I = 10^{-39} - 10^{-38} g \cdot cm^2$ (heavy molecules, $10^4 - 10^5$ electron mass for the molecular mass, $10^{-8} cm$ for the dipole length) we get $\Omega \simeq 10^{11} s^{-1}$; this is the lower limit of rotation molecular spectra (the upper limit of vibrational molecular spectra goes up to $10^{14} s^{-1}$, both are in the infrared range).[7]

For high-intensity electric fields, *i.e.* for $dE \gg M_z^2/I \simeq T$, which implies fields of the order $E \simeq 10^4 statvolt/cm$, for $d = 10^{-18} esu$ and temperature $T = 300K$ (for comparison, the atomic fields are of the order $-e/a_H^2$, which typically is $\simeq 10^6 statvolt/cm$), the potential $V_{eff} + dE\cos\theta$ has a minimum for $\theta_0 = (M_z^2/IdE)^{1/4} \ll 1$; the motion is a θ-oscillation and a φ-rotation with frequency $\omega_0 = 2\sqrt{dE/I}$; this frequency is known as the Rabi frequency;[8] with our numerical data it is much higher than $10^{12} s^{-1}$. Such high–intensity fields may occur in condensed matter from dipole-dipole interaction. These electric fields are very similar with those which may act upon rotating molecules in crystals, driving the transition from rotations to vibrations (quenched rotations) which is seen in the curve of the heat capacity *vs* temperature.[9]

If we limit the rotation to a plane, then the lagrangian

$$L = \frac{1}{2} I\dot{\varphi}^2 + dE\cos\varphi = \frac{1}{2I} M^2 + dE\cos\theta \tag{3.139}$$

is the lagrangian of the planar rotator; for weak field the motion is an oscillation about $\varphi = 0$ with Rabi's frequency $\omega_0 = \sqrt{dE/I}$.

In general, the rotation motion of the dipoles is complicated; the angular variables θ, φ are not proper dynamical variables, since they are multi-valued functions of the motion.

The quadrupole moment

$$Q_{ij} = \sum_a q_a (3u_{ai}u_{aj} - \delta_{ij}u_a^2) \tag{3.140}$$

[7] M. Apostol and L. C. Cune, "Molecular dynamics in high electric fields", Chem. Phys. **472** 262 (2016).

[8] I. I. Rabi, "On the process of space quantization", Phys. Rev. **49** 324 (1936); I. I. Rabi, "Space quantization in a gyrating magnetic field", Phys. Rev. **51** 652 (1937).

[9] L. Pauling, "The rotational motion of molecules in crystals", Phys. Rev. **36** 430 (1930); T. E. Stern, "The symmetrical spherical oscillator, and the rotational motion of homopolar molecules in crystals", Proc. Roy. Soc. **A130** 551 (1931).

can be brought to its principal axes; since it is traceless, only two parameters characterize this tensor. Its motion is that of a rigid body. We may give up the motion of individual particles, and average the quadrupole moment over this motion; then, it will be expressed by means of the only vector available for the assembly of the particles, which is the angular momentum; therefore, we have

$$Q_{ij} = \frac{3Q}{2M^2}\left(M_i M_j - \frac{1}{3}M^2\delta_{ij}\right) \ , \tag{3.141}$$

where the constant Q is chosen such that $Q_{zz}(M_z = M) = Q$. The motion of Q_{ij} given by equation (3.141) is the motion of the angular momentum under the action of the torque of the forces.

3.10.5 Time-dependent uniform electric field

Let us consider a component $\mathbf{A} = \mathbf{A}_0 \cos(\omega t - \mathbf{kr})$ of the vector potential of the electromagnetic radiation, where ω is the frequency and $\mathbf{k}$ is the wavevector, $\omega = ck$. Since the velocity of a non-relativistic particle is much smaller than the speed of light in vacuum c, we may neglect the spatial phase $\mathbf{kr}$ and write the vector potential as $\mathbf{A} = \mathbf{A}_0 \cos\omega t$; this amounts to neglecting the magnetic field; then, the lagrangian of a charge q with mass m is

$$L = \frac{1}{2}mv^2 + q\frac{\mathbf{v}}{c}\mathbf{A} \tag{3.142}$$

and the standard non-relativistic hamiltonian is

$$H = \frac{1}{2m}(\mathbf{p} - \frac{q}{c}\mathbf{A})^2 \ , \tag{3.143}$$

with momentum $\mathbf{p} = m\mathbf{v} + q\mathbf{A}/c$. If we use $\mathbf{v} = d\mathbf{r}/dt$ and a total time derivative, the lagrangian becomes

$$L = \frac{1}{2}mv^2 + q\mathbf{rE} \tag{3.144}$$

and the hamiltonian is

$$H = \frac{1}{2m}p^2 - q\mathbf{rE} \ , \tag{3.145}$$

with momentum $\mathbf{p} = m\mathbf{v}$, where $\mathbf{E} = -(1/c)\partial\mathbf{A}/\partial t = -(1/c)d\mathbf{A}/dt$ is the electric field; these are the dipole lagrangian and hamiltonian; we note that the dipole in these equations is an orbital dipole. We note that the hamiltonian given by equation (3.143) corresponds to a charge immersed in the electromagnetic field, while in the hamiltonian given by equation (3.145) the electromagnetic field may be restricted to a finite spatial region.

3.10.6 Uniform, constant magnetic field. Magnetic moments

Let us consider a uniform magnetic field $\mathbf{H}$, which may depend on the time; it derives from a vector potential $\mathbf{A} = \frac{1}{2}\mathbf{H} \times \mathbf{r}$, with $div\mathbf{A} = 0$. The lagrangian of a non-relativistic charge q with mass m in this field is

$$\begin{aligned} L &= \tfrac{1}{2}mv^2 + q\tfrac{\mathbf{v}}{c}\mathbf{A} = \tfrac{1}{2}mv^2 + q\tfrac{\mathbf{v}}{2c}(\mathbf{H} \times \mathbf{r}) = \\ &= \tfrac{1}{2}mv^2 + \boldsymbol{\mu}\mathbf{H} \ , \end{aligned} \tag{3.146}$$

where

$$\boldsymbol{\mu} = \frac{q}{2c}(\mathbf{r} \times \mathbf{v}) = \frac{q}{2mc}(\mathbf{r} \times \mathbf{p}) = \frac{q}{2mc}\mathbf{m} \tag{3.147}$$

is the magnetic moment (magnetic dipole); $\mathbf{p} = m\mathbf{v}$ is the momentum and $\mathbf{m} = \mathbf{r} \times \mathbf{p}$ is the angular momentum; $q/2mc$ is called the gyromagnetic factor. We note that $\boldsymbol{\mu}$ is an orbital magnetic moment. The hamiltonian is

$$\begin{aligned} H &= \tfrac{1}{2m}(\mathbf{p} - \tfrac{q}{c}\mathbf{A})^2 = \tfrac{1}{2m}p^2 - \tfrac{q}{mc}\mathbf{p}\mathbf{A} + \tfrac{q^2}{2mc^2}A^2 = \\ &= \tfrac{1}{2m}p^2 - \boldsymbol{\mu}\mathbf{H} + \tfrac{q^2}{8mc^2}[r^2H^2 - (\mathbf{r}\mathbf{H})^2] \ , \end{aligned} \tag{3.148}$$

where $\mathbf{p} = m\mathbf{v} - q\mathbf{A}/c$. The term $-\boldsymbol{\mu}\mathbf{H}$ tends to allign the magnetic moment along the field; it is the paramagnetic term; on the contrary, the term quadratic in H in equation (3.148) tends to move the charge towards regions where the field is weaker; it is the diamagnetic term. For an assembly of charges with the same ratio q/m, the magnetic moment is

$$\boldsymbol{\mu} = \frac{q}{2mc}\sum_a m_a(\mathbf{r} \times \mathbf{v}_a) = \frac{q}{2mc}(\mathbf{r} \times \mathbf{P}) = \frac{q}{2mc}\mathbf{M} \ , \tag{3.149}$$

where **P** is the total momentum and **M** is the total angular momentum.

The total orbital moments are zero, in many cases; they are quenched, due to equal distribution of velocities $\mathbf{v}$ and $-\mathbf{v}$; for example, the statistical average of the orbital moment with classical distribution is zero, since the classical distribution is quadratic in velocities (this is known as the Bohr-van Leuween theorem.[10]); similar quenching of magnetic moments occurs in centrally symmetric potentials, as a consequence of the parity invariance.

We may envisage composite particles, consisting of many charges with a finite motion; then, we may average over the finite motion; this way, we get composite particles with orbital moments. For instance, since $\frac{1}{2}\mathbf{r}\times\mathbf{v} = dS/dt$ is the area in unit time, it follows $\boldsymbol{\mu} = qd\mathbf{S}/cdt$; for a uniform rotation $\mu = IS/c$, where $I = q/T = \nu q$ is the current, ν is the frequency and T is the rotation period, $S = \pi r^2$ is the area enclosed by the circular orbit; we have $\mu = (q/2c)\omega r^2$, the angular momentum $m = m\omega r^2$ and $\mu = (q/2mc)m$, where $\omega = 2\pi\nu$ is the angular frequency. We can see that the magnetic dipole can be visualized as a small coil carrying an electric current. The diamagnetic term in equation (3.148) can be averaged over the finite motion; we get an energy $\Delta E = (q^2/12mc^2)\overline{r^2}H^2$, a magnetization $-\partial\Delta E/\partial H = -(q^2/6mc^2)\overline{r^2}H$ and a diamagnetic susceptibility $\chi_d = -(q^2/6mc^2)\overline{r^2}$ (this is known as the Langevin diamagnetism); similarly, we can define the paramagnetic magnetization and the paramagnetic susceptibility (the Curie-Langevin-Debye magnetization).

Let a particle with internal velocity $\mathbf{v}_a$ be placed at $\mathbf{r}_a$; it carries a current density $\mathbf{j}_a = q\mathbf{v}_a\delta(\mathbf{r}-\mathbf{r}_a)$; it generates a vector potential $\mathbf{a}$ given by the wave equation

$$\frac{1}{c^2}\ddot{\mathbf{a}} - \Delta\mathbf{a} = \frac{4\pi}{c}\mathbf{j}_a \ ; \tag{3.150}$$

the solution of this equation is

$$\mathbf{a} = \frac{1}{c}\int d\mathbf{r}' \frac{\mathbf{j}_a(\mathbf{r}', t - \left|\mathbf{r}-\mathbf{r}'\right|/c)}{|\mathbf{r}-\mathbf{r}'|} \ ; \tag{3.151}$$

[10] N. Bohr, Dissertation, Copenhagen (1911); J. H. van Leeuwen, Dissertation, Leiden (1919); J. H. van Vleck, *Theory of Electric and Magnetic Susceptibilities*, Oxford (1932).

we may neglect the retardation for non-relativistic particles; or, we may average over the finite motion, and equation (3.150) becomes

$$\Delta \mathbf{a} = -\frac{4\pi}{c}\mathbf{j}_a \tag{3.152}$$

with the solution

$$\mathbf{a} = \frac{1}{c}\int d\mathbf{r}' \frac{\mathbf{j}_a(\mathbf{r}')}{|\mathbf{r}-\mathbf{r}'|} = \frac{q}{c}\frac{\mathbf{v}_a}{|\mathbf{r}-\mathbf{r}_a|} \ ; \tag{3.153}$$

this is the Biot-Savart law; the Lorenz gauge $(1/c)\partial\varphi/\partial t + div\mathbf{a} = 0$, where φ is the scalar protential, is satisfied by the continuity equation. We expand in powers of $\mathbf{r}_a$,

$$\mathbf{a} = \frac{q}{c}\frac{\mathbf{v}_a}{r} + \frac{q}{c}\frac{\mathbf{v}_a(\mathbf{r}\mathbf{r}_a)}{r^3} + ... \ ; \tag{3.154}$$

the first term is zero on averaging over the finite motion, while the second can be written as

$$\mathbf{v}_a(\mathbf{r}_a\mathbf{r}) = \frac{1}{2}\frac{d}{dt}[\mathbf{r}_a(\mathbf{r}_a\mathbf{r})] + \frac{1}{2}[\mathbf{v}_a(\mathbf{r}_a\mathbf{r}) - \mathbf{r}_a(\mathbf{v}_a\mathbf{r})] \ , \tag{3.155}$$

or, by a further averaging,

$$\mathbf{v}_a(\mathbf{r}_a\mathbf{r}) \rightarrow \frac{1}{2}[\mathbf{v}_a(\mathbf{r}_a\mathbf{r}) - \mathbf{r}_a(\mathbf{v}_a\mathbf{r})] = \frac{1}{2}(\mathbf{r}_a \times \mathbf{v}_a) \times \mathbf{r} \ ; \tag{3.156}$$

the potential vector given by equation (3.154) becomes

$$\mathbf{a} = \frac{q}{2c}\frac{(\mathbf{r}_a \times \mathbf{v}_a)\times \mathbf{r}}{r^3} = \frac{\boldsymbol{\mu}\times\mathbf{r}}{r^3} \tag{3.157}$$

for large distances. The magnetic field $\mathbf{h} = curl\mathbf{a}$ is

$$\mathbf{h} = \frac{3(\boldsymbol{\mu}\mathbf{r})\mathbf{r} - \boldsymbol{\mu}r^2}{r^5} \tag{3.158}$$

(where a $\Delta(1/r) = -4\pi\delta(\mathbf{r})$ was left aside). This is the magnetic field of the magnetic dipole; the interaction energy of two magnetic dipoles $\boldsymbol{\mu}_{1,2}$ is

$$\Delta E = -\boldsymbol{\mu}_1\mathbf{h}_2 = \frac{\boldsymbol{\mu}_1\boldsymbol{\mu}_2 r^2 - 3(\boldsymbol{\mu}_1\mathbf{r})(\boldsymbol{\mu}_2\mathbf{r})}{r^5} \tag{3.159}$$

(magnetic dipole interaction).

The force

$$\mathbf{F} = \sum_a \frac{q_a}{c} \mathbf{v}_a \times \mathbf{H} \tag{3.160}$$

acting upon an assembly of moving charges placed in a magnetic field $\mathbf{H}$ is zero when averaged over the finite motion and the field is constant; the torque is given by

$$\begin{gathered}\mathbf{K} = \sum_a \frac{q_a}{c}[\mathbf{r}_a \times (\mathbf{v}_a \times \mathbf{H})] = \\ = \sum_a \frac{q_a}{c}\left[\mathbf{v}_a(\mathbf{r}_a\mathbf{H}) - \frac{1}{2}\mathbf{H}\frac{d}{dt}(r_a^2)\right] \rightarrow \\ \rightarrow \sum_a \frac{q_a}{c}\mathbf{v}_a(\mathbf{r}_a\mathbf{H}) \ ,\end{gathered} \tag{3.161}$$

or

$$\begin{gathered}\mathbf{K} = \sum_a \frac{q_a}{2c}[\mathbf{v}_a(\mathbf{r}_a\mathbf{H}) - \mathbf{r}_a(\mathbf{v}_a\mathbf{H})] = \\ = \sum_a \frac{q_a}{2c}[\mathbf{H} \times (\mathbf{v}_a \times \mathbf{r}_a)] = \boldsymbol{\mu} \times \mathbf{H} \ ,\end{gathered} \tag{3.162}$$

by using the same averaging procedure as above. It follows that the equation of motion of the magnetic moment in a constant magnetic field is

$$\frac{d\boldsymbol{\mu}}{dt} = \frac{q}{2mc}\frac{d\mathbf{M}}{dt} = \frac{q}{2mc}\boldsymbol{\mu} \times \mathbf{H} \ . \tag{3.163}$$

This is the Larmor equation of motion (precession equation).

Let us assume an assembly of charges with the lagrangian

$$L = \sum_a \frac{1}{2} m_a v_a^2 - U \tag{3.164}$$

with usual notations, where U is their potential energy (including a centrally symmetric field and interaction). In a frame rotating with angular velocity $\boldsymbol{\Omega}$ the velocity is given by

$$\mathbf{v} = \mathbf{v}^{'} + \boldsymbol{\Omega} \times \mathbf{r}^{'} \ , \tag{3.165}$$

while the potential energy does not change. The lagrangian becomes

$$L = \sum_a [\frac{1}{2} m_a v_a^{'2} + m_a \boldsymbol{\Omega}(\mathbf{r}_a^{'} \times \mathbf{v}_a^{'}) + \frac{1}{2} m_a (\boldsymbol{\Omega} \times \mathbf{r}_a^{'})^2] - U \ ; \tag{3.166}$$

for the same ratio charge-to-mass (q/m) and for $\boldsymbol{\Omega} = \frac{q}{2mc}\mathbf{H}$ we can see that the lagrangian acquires a magnetic term $\boldsymbol{\mu}\mathbf{H}$ (on averaging over the finite microscopic motion of charges), providing the magnetic field $\mathbf{H}$ (and angular velocity $\boldsymbol{\Omega}$) are sufficiently small as to neglect the quadratic term in H^2 (Ω^2). This is known as Larmor's theorem; the angular velocity $\Omega = |q|\,H/2mc$ is called the Larmor frequency. The motion of the magnetic and angular moments is governed by the angular velocity $\boldsymbol{\Omega}$. Indeed,

$$\frac{d\mathbf{M}}{dt} = -\boldsymbol{\Omega} \times \mathbf{M} \tag{3.167}$$

(in the rotating frame), whence

$$\frac{d\boldsymbol{\mu}}{dt} = \frac{q}{2mc}\boldsymbol{\mu}\times\mathbf{H} \ . \tag{3.168}$$

It is worth noting that the motion of a charge q in a constant magnetic field H proceeds according to the equations $m\dot{v}_x = \frac{q}{c}v_y H$, $m\dot{v}_y = -\frac{q}{c}v_x H$, *i.e.* $\ddot{v}_x = (qH/mc)^2 v_x$; this motion oscillates with the frequency $\frac{qH}{mc}$, which is known as the cyclotron frequency. It is the average over microscopic motion which makes the magnetic moment and the angular moment to precess with Larmor's frequency $\frac{qH}{2mc}$.

Apart from their orbital magnetic moments, related to the orbital angular momenta, the particles may have intrinsic magnetic moments, related to their spins (intrinsic angular momenta). For an orbit with radius r the momentum is quantized as $p = (2\pi\hbar/2\pi r)n = (\hbar/r)n$, where n is any (non-vanishing) integer; the spins can be half-integers too; all the known particles have spin zero or 1/2; the quantum of the angular momentum is $\Delta M = \hbar$; it follows that we can write the magnetic moments as $\mu = (q\hbar/2mc)M$, where M stands now for the orbital angular momentum L, or spin S, or total angular momentum $\mathbf{L} = \mathbf{L} + \mathbf{S}$, all measured in pure numbers; lower-case letters can also be used, like l, s, j. For electron $\mu_B =\mid e \mid \hbar/2mc \simeq 10^{-20} erg/Gs$ is called the Bohr magneton; $e = -4.8 \times 10^{-10} esu$ is the electron charge and $m \simeq 10^{-27} g$ is the electron mass; $\hbar \simeq 10^{-27} erg \cdot s$ is Planck's constant. The Bohr magneton is used as the unit for atomic magnetic moments. A Bohr magneton generates a dipole magnetic field $\mu_B/r^3 \simeq 10^4 Gs$ ($1Gs = 10^{-4}Ts$) at distance $r = 1\text{Å}$; this is a rather high magnetic field, in comparison with the usual magnetic fields in the laboratory.[11]

In many-particle assemblies each particle may be viewed as moving in a common, self-consistent mean-field, which depends on the assembly of the particles. The mean-field concept, introduced by Bohr, is helpful in atoms, molecules, atomic nuclei. Usually, the single-(one)-particle states are grouped in shells, with a finite number of states. Fully occupied shells (closed shells) have zero angular momentum and zero spin; therefore, they have zero magnetic moments. Incomplete shells (open shells) have angular momenta and spins and, consequently, magnetic moments. In molecules, or in condensed matter, the orbital momentum is quenched (is zero); usually, molecules have zero spin in their ground-state. For nuclear magnetic moments the unit is the nuclear Bohr magneton, which is the atomic Bohr magneton with the electron mass replaced by the proton mass; it follows that the nuclear magneton is $\simeq 5 \times 10^{-24} erg/Gs$. In general, interactions change the gyromagnetic factor, so that the magnetic moment is written as $\boldsymbol{\mu} = g\mu_B L$ (or S, or J), where the factor g is sometimes called the Lande factor. Deformation brought about by rotation can induce magnetic moments (and quadrupole electric moments). Hyperfine interaction arises from the interaction of the nucleus with the surrounding electrons. Matrix elements of magnetic moments between ground-state and excited states are, in general, non-zero. External fields may induce magnetic moments.

[11] More details about electric and magnetic moments in M. Apostol, *Magnetic and Electric Resonance*, Cambridge Scholars Publishing, Cambridge (2018).

4 Quantum Mechanics

4.1 Heisenberg's matricial mechanics

Motion is described by functions of time; any such function is decomposable in a Fourier series,

$$F(t) = \frac{1}{2\pi}\int d\omega e^{-i\omega t}F(\omega) \ ; \tag{4.1}$$

according to the quantization idea, we can replace ω in this equation by $\omega = E/\hbar$, where $\hbar = h/2\pi$, h and $\hbar$ being Planck's constants. However, these energies are discrete, then we should have $\omega_n = E_n/\hbar$, where n is an integer, and the integral should be replaced by a sum. We note that discrete frequencies appear naturally in Fourier series of periodic functions, *i.e.* for periodic motion. However, we measure only energy differences, such that we should have

$$e^{-\frac{i}{\hbar}(E_m-E_n)t} \tag{4.2}$$

instead of $e^{-i\omega t}$, where n, m denote sets of integers (the order of $E_{n,m}$ is arbitrary); in this case $F(\omega)$ should be replaced by a quantity denoted F_{nm}, such that

$$F(t) = \sum_{n,m} e^{-\frac{i}{\hbar}(E_m-E_n)t}F_{nm} \ ; \tag{4.3}$$

this was the starting point of Heisenberg's matricial mechanics.[1] The representation given by equation (4.2) is a new representation for time-dependent functions, with far-reaching implications. We note

[1]W. Heisenberg, "Uber quantentheoretische Umdeutung kinematischer und mechanischer Beziehungen", Z. Phys. **33** 879 (1925).

the principle that "only measurable quantities are admitted in theory" (which, in fact, is a relative principle) and the additional freedom brought by two labels instead of one in equation (4.3).

The quantities F_{nm} were recognised to be matrix elements.[2] It implies that an operator F is associated to any physical quantity; such quantum-mechanical quantities are called sometimes q-numbers, in order to distinguish them from ordinary complex numbers, which are called c-numbers. F acts in a linear space of vectors φ with a scalar product (Hilbert space) of the form (φ, ψ); the operator F has eigenvectors φ_n and eigenvalues f_n, such that

$$F\varphi_n = f_n\varphi_n \ ; \tag{4.4}$$

the matrix elements are written as

$$F_{nm} = (\varphi_n, F\varphi_m) \ , \tag{4.5}$$

and the mean value is $F_{nn} = (\varphi_n, F\varphi_n)$, or, generally, $\overline{F} = (\varphi, F\varphi)$. It is assumed that the mean values are measurable quantities; it is also assumed that the eigenvalue f_n is the value the quantity F has in the state φ_n; since they are real, we should have

$$f_n = (\varphi_n, F\varphi_n) = (\varphi_n, F\varphi_n)^* = (F\varphi_n, \varphi_n) = (\varphi_n, F^{t*}\varphi_n) \ , \tag{4.6}$$

where F^t is the transposed operator, F^* is the complex-conjugate operator, $F^{t*} = F^+$ is the transposed and conjugate, or adjoint, operator; it follows that the operator F must have the property $F = F^+$; we say that F must be self-adjoint, or hermitian. A hermitian operator has orthogonal eigenvectors for distinct eigenvalues, since

$$\begin{gathered}(\varphi_n, F\varphi_m) = f_m(\varphi_n, \varphi_m) \ , \\ (\varphi_n, F\varphi_m) = (F\varphi_n, \varphi_m) = f_n(\varphi_n, \varphi_m)\end{gathered} \tag{4.7}$$

and

$$(f_n - f_m)(\varphi_n, \varphi_m) = 0 \ ; \tag{4.8}$$

[2]M. Born and P. Jordan, "Zur Quantenmechanik", Z. Phys. **34** 858 (1925); M. Born, W. Heisenberg and P. Jordan, "Zur Quantenmechanik II", Z. Phys. **35** 557 (1926).

if there exist several eigenvectors for the same eigenvalue, that eigenvalue is degenerate, and we can orthogonalize the degenerated eigenvectors; we may also orthonormalize the eigenvectors, *i.e.* we can impose the condition $(\varphi_n, \varphi_m) = \delta_{nm}$. The set of eigenvalues f_n is the spectrum of F (together with the corresponding eigenvectors); it may be discrete, or continuous; in the latter case the label n is continuous and we may write f_n as $f(n)$ and sums of the type $\sum_n$ should be replaced by $\int dn$; the orthonormality condition reads in that case $(\varphi_n, \varphi_m) = \delta(n-m)$, where δ is Dirac's delta function. By $\sum$ or $\int$ we understand indistinctly either $\sum$, or $\int$, or both.

Since the eigenvectors are linearly independent (and orthogonal), they may be taken as a basis of expansion for any other vector

$$\psi = \sum_n c_n \varphi_n \quad , \tag{4.9}$$

where c_n are some complex coefficients. It follows that the vectors represent the states of the "physical system", and any state can be written as a superposition of other states; the operators act on the state vectors; the eigenvectors are also called eigenstates. This is the superposition principle of the Quantum Mechanics. The mean value is

$$\overline{F} = (\psi, F\psi) = \sum_n f_n \mid c_n \mid^2 \quad , \tag{4.10}$$

which means that $\mid c_n \mid^2$ is the probability to have the state φ_n in the state ψ, and $c_n = (\varphi_n, \psi)$ is the amplitude of the corresponding probability. We must have

$$(\psi, \psi) = \mid \psi \mid^2 = \sum_n \mid c_n \mid^2 = 1 \tag{4.11}$$

(since φ_n are orthonormal); $\mid \psi \mid$ is the length of the vector ψ. Similarly, φ_n is the amplitude of probability of being in state n (or φ_n), and $(\varphi_n, \varphi_n) = \mid \varphi_n \mid^2 = 1$ is the probability of being in state n, which is unity.[3] We can see that the vector ψ defines the state up to a phase. We note also that the quantum-mechanical operators are linear operators. The expansion given by equation (4.9) means that

[3] M. Born, "Zur Quantenmechanik der Stossvorgange", Z. Phys. **37** 863 (1926).

the eigenvectors φ_n form a complete (or closed) set of vectors, and equation (4.11) is the closure equation.

The matrices (operators) do not commute with one another, in general; we write

$$FG - GF = [F, G] \neq 0 \tag{4.12}$$

for two operators F and G that do not commute; $[F, G]$ is called the commutator of F and G (or the commutator of F with G). It is easy to see that if two operators commute, they have common eigenvectors, and if they have common eigenvectors, they commute. This means that they have well-defined values on the same state, they may be simultaneously measured. It follows that the quantum-mechanical state is defined as the common eigenvector of the maximum number of commuting operators. This would be the maximum information we may have about the quantum-mechanical motion. The hermitian operators associated to physical quantities are also called observables.

A matrix (operator) can be diagonalized, in general; hermitian operators always.[4] The linear transform U which diagonalizes an operator F acts like

$$U^{-1}FU = \widetilde{F} \quad , \tag{4.13}$$

where $\widetilde{F}$ is diagonal; if F is hermitian, then $U^{-1} = U^{+}$ and $U^{+}U = 1$; then, U is called an unitary operator; it may be written as $U = e^{iS}$, where S is hermitian. A unitary transformation preserves the commutation relations.

A convenient notation has been introduced by Dirac, who has written the scalar product as a "bracket" $(\varphi, \psi) =< \varphi \mid \psi >$; then, Dirac called "kets" the vectors $\mid \psi >$ and "bras" the vectors $< \varphi \mid$.[5] Moreover, Dirac viewed functions of the form φ_n, or $\varphi(n)$ if n is continuous, as scalar products $< n \mid \varphi >$; the closure equation (4.11) can be written as

$$\begin{gathered}(\psi, \psi) = \sum_n \mid c_n \mid^2 = \sum_n (c, n)(n, c) = \\ = \sum_n < c \mid n >< n \mid c >= 1 \quad ,\end{gathered} \tag{4.14}$$

[4]The necessary and sufficient condition for a matrix to be diagonalizable is to commute with its transposed (then it is called a normal matrix).

[5]P. A. M. Dirac, *The Principles of Quantum Mechanics*, Oxford University Press, Oxford (1930).

whence

$$\sum_n \mid n >< n = 1 \quad , \tag{4.15}$$

since $< c \mid c >= 1$.

4.2 Equations of motion

Differentiating equation (4.3) with respect to the time we get

$$\frac{d}{dt} F_{nm} = \frac{i}{\hbar}(E_n - E_m) F_{nm} \quad . \tag{4.16}$$

We may imagine that the states n are eigenvectors of the hamiltonian H,

$$H \varphi_n = E_n \varphi_n \tag{4.17}$$

(or $H \mid n >= E_n \mid n >$); then, removing the scalar products in equation (4.16), we get

$$\frac{dF}{dt} = \frac{i}{\hbar}(HF - FH) = \frac{i}{\hbar}[H, F] \quad ; \tag{4.18}$$

this equation is general; it can also be written as

$$F(t) = e^{\frac{i}{\hbar}Ht} F(0) e^{-\frac{i}{\hbar}Ht} \quad . \tag{4.19}$$

The resemblance of the equation (4.18) with the classical equation of motion

$$\frac{dF}{dt} = \{H, F\} \quad , \tag{4.20}$$

where $\{\}$ is the Poisson bracket, is striking; to the same extent to which equation (4.20) gives the equations of motion of Newton's Mechanics, equation (4.18) gives the quantum-mechanical equations of motions. The quantum-mechanical equations of motion are given by the commutator with the hamiltonian. These are the canonical equations of motion; for this reason, the unitary transformations which diagonalize hermitian operators (the observables), and preserve the commutation relations, are also called canonical transformations.

The association

$$\frac{i}{\hbar}[H,F] \to \{H,F\} \ , \ \ [H,F] = -i\hbar\{H,F\} \tag{4.21}$$

was noted by Dirac;[6] it opened the way to the new Quantum Mechanics. It is worth noting that in the limit $\hbar \to 0$ the commutator is vanishing: the operators become classical quantities. If the operators have an intrinsic time dependence, not given by the dynamics, then their full equation of motion reads

$$\frac{dF}{dt} = \frac{\partial F}{\partial t} + \frac{i}{\hbar}[H,F] \ , \tag{4.22}$$

in full analogy with the classical equations of motion.
In scalar products of the form

$$(\varphi, F(t)\psi) = (\varphi, e^{\frac{i}{\hbar}Ht}F(0)e^{-\frac{i}{\hbar}Ht}\psi) \tag{4.23}$$

we may write

$$\begin{aligned}(\varphi, F(t)\psi) &= (\varphi, e^{\frac{i}{\hbar}Ht}F(0)e^{-\frac{i}{\hbar}Ht}\psi) = \\ &= (e^{-\frac{i}{\hbar}Ht}\varphi, F(0)e^{-\frac{i}{\hbar}Ht}\psi)\end{aligned} \tag{4.24}$$

and we may imagine that the operators do not depend on time but, instead, the state vectors do; then we write simply $F(0) = F$ and

$$\varphi(t) = e^{-\frac{i}{\hbar}Ht}\varphi \ ; \tag{4.25}$$

which leads to the equation

$$i\hbar\frac{\partial}{\partial t}\varphi(t) = H\varphi(t) \ ; \tag{4.26}$$

this equation was later called the Schrodinger equation. Time-independent operators and time-dependent state vectors of the form given by equation (4.25) are said to belong to the Schrodinger picture (or

[6] P. A. M. Dirac, "The fundamental equations of Quantum Mechanics", Proc. Roy. Soc. **A109** 642 (1926).

Schrodinger representation); while time-dependent operators and time-independent state vectors of the form given by equation (4.23) are said to be in the Heisenberg picture (or Heisenberg representation). If the Heisenberg picture is performed only for the non-interacting part of the hamiltonian, while the interaction is left in the Schrodinger picture, then we say that we have the interaction picture.

We note here that in the above considerations H is independent of time, as for a closed, conservative motion, and the (constant) energies do exist. The corresponding states are called stationary states. If the hamiltonian depends on the time, then the energy eigenvalues $E_n(t)$ and the corresponding eigenvectors $\varphi_n(t)$ depend on the time; although, formally we may define time-dependent frequencies $\omega(t) = [E_n(t) - E_m(t)]/h$, the quantization idea of absorbing and emitting such energy differences raises difficulties, since the absorption and emission processes are not instantaneous. However, the Poisson brackets for the classical dynamics are valid, and their quantum-mechanical counterpart with commutators is their expression for finite amounts h of mechanical action; the classical limit is achieved by these equations of motion in the limit $h \to 0$, so we may maintain the quantum-mechanical equations of motion for time-dependent hamiltonians; the only difference is that expressions like Ht (in equation (4.19), for instance) are replaced by $\int^t dt' H(t')$ and expressions like $E_n t$ (in equation (4.3), for instance) are replaced by $\int^t dt' E_n(t')$. In addition, the time-dependent Schrodinger equation is satisfied by a linear combination of wavefunctions $\varphi_n(t)$ with time-dependent coefficients.

Let us consider a particle with mass m in a potential $V(\mathbf{r})$; the hamiltonian of its motion is

$$H = \frac{1}{2m}p^2 + V(\mathbf{r}) \quad , \tag{4.27}$$

where $\mathbf{p}$ is momentum. The motion of both the position $\mathbf{r}$ and the momentum $\mathbf{p}$ is given by

$$\begin{gathered}\dot{\mathbf{r}} = \frac{i}{\hbar}[H, \mathbf{r}] = \frac{i}{\hbar}[\frac{1}{2m}p^2, \mathbf{r}] = \frac{i}{\hbar}\frac{1}{2m}\{\mathbf{p}[\mathbf{p}, \mathbf{r}] + [\mathbf{p}, \mathbf{r}]\mathbf{p}\} \ , \\ \dot{\mathbf{p}} = \frac{i}{\hbar}[\mathbf{H}, \mathbf{p}] = \frac{i}{\hbar}[V(\mathbf{r}), \mathbf{p}] \ ;\end{gathered} \tag{4.28}$$

if in the classical limit $\hbar \to 0$ we are going to recover the Newton's equations of motion

$$\dot{\mathbf{r}} = \mathbf{p}/m \ , \ \dot{\mathbf{p}} = -\frac{\partial V}{\partial \mathbf{r}} \tag{4.29}$$

from the above equations (according to the correspondence principle), we should have

$$[\mathbf{p}, \mathbf{r}] = -i\hbar \ , \ [p_x, x] = -i\hbar \ , \ [p_x, y] = 0 \ , etc \ ; \tag{4.30}$$

these commutation relations should be read $[p_i, x_j] = -i\hbar\delta_{ij}$, where i, j denote cartesian components. The first Newton's equation follows immediately from the first equation (4.28) and the second Newton's equation follows from the second equation (4.28) by a series expansion. It follows that the commutation relations between canonical conjugate variables are the fundamental elements of the quantum-mechanical dynamics.

4.3 Uncertainty principle

The eigenvalue f_n in the eigenvalues equation $F\varphi_n = f_n\varphi_n$ (or $F \mid n >= f_n \mid n >$) is the quantity F we are able to measure (exactly) in state n. We can also measure the mean value $\overline{F}$ for a state φ, which is not an eigenstate of F, but, as any mean value, this is a statistical value; which means that in state φ the quantity F has a statistical deviation δF; this is defined as

$$\delta F = \sqrt{\overline{F^2} - \overline{F}^2} \ ; \tag{4.31}$$

for a general state n the mean value of F is $\overline{F} = F_{nn}$ and the mean square value is

$$\overline{F^2} = \sum_m F_{nm}F_{mn} = \sum_m |F_{nm}|^2 \ ; \tag{4.32}$$

we can see that $\overline{F^2}$ and the deviation δF are related to the off-diagonal matrix elements; and

$$\delta F = \sqrt{\sum_m{}' |F_{nm}|^2} \ , \tag{4.33}$$

where the prime on the summation sign means $m \neq n$.

If two operators do not commute they cannot be measured simultaneously, because they have not common eigenvectors; it follows that they have statistical deviations. These deviations can be measured, because we can measure mean values. How, in what sense, can we do that? We may imagine statistical ensembles of the same "physical system", *i.e.* either we prepare the system many times in the same conditions, or prepare many systems in the same conditions; then, we proceed to measure, say, the quantity F; we shall get values f_n with frequency w_n; the mean value will be $\overline{F} = \sum_n f_n w_n$; then we will know that the system is in state $\varphi = \sum_n c_n \varphi_n$, where $F\varphi_n = f_n \varphi_n$ and $w_n = \mid c_n \mid^2$. We will also know that each time we measure F we force the system to be in the eigenstate φ_n, *i.e.* each time we make a measurement the state vector of the system φ becomes the eigenvector φ_n; each time we make a measurement the state vector of the system is destroyed, and is replaced by φ_n. This is called the collapse of the state vector, produced by the measurement process. The reason for this circumstance arises from the fact that the discrete quantum quantities imply the Planck's constant h, which at macroscopic scale is very small; therefore, during measurement, which is inevitably made by means of a macroscopic apparatus, we disturb appreciably the system, and force it to be in the macroscopically measurable state φ_n. We can see that the measuring subject, *i.e.* the human being, intervenes essentially in the dynamic behaviour of the quantum systems by the measuring apparatus, though it is not our subjectivity which dictates the measurement results, but the statistical nature of the measuring act. This is the so-called statistical interpretation (meaning) of the Quantum Mechanics, or Copenhagen interpretation of the Quantum Mechanics; it was put forward by Bohr in Copenhagen and his collaborators. In this context, the separate measurement of two non-commuting operators is called "complementarity".

In the measurement process described above the mean square of the quantity F is $\overline{F^2} = \sum_n f_n^2 w_n$, where $w_n = \mid c_n \mid^2$; the quantity F has a statistical deviation. The system has a state vector $\varphi = \sum_n c_n \varphi_n$ in this case. In general, φ_n are not necessarily the eigenvectors of F; then, the mean value of F is $\overline{F} = \sum_{nm} c_n^* c_m F_{nm}$ (and a similar expression for $\overline{F^2}$); we may view $c_n^* c_m$ as the matrix element $w_{mn} = c_n^* c_m$ of some matrix (operator) w; then $\overline{F} = \sum_m (wF)_{mm} = tr(wF)$, where tr is the notation for the trace (die Spur) of the matrix; it is called the density matrix;[7] we note that $tr(w) = 1$ and $w^2 = w$. If the density matrix is given by a state vector, *i.e.* if $w_{mn} = c_n^* c_m$, the system is said to be a pure system (or to be in pure state); it is describable by a state vector, its state is "well defined" (quantum-mechanically). We may imagine that the system is not well (completely) describable by itself, it may depend on other systems for instance; then, the density matrix w_{nm} is not decomposable as a product $c_n^* c_m$, but the mean value is still definable as $\overline{F} = \sum_{nm} w_{mn} F_{nm} = tr(wF)$; in this case the system is said to be in a mixed state. We note that the property $trw = 1$ is preserved, but not the idempotency property ($w^2 \neq w$). The time-dependence of the operators F can be transferred to the density matrix w; then, it is worth noting that the equation of motion of the density matrix is

$$\dot{w} = \frac{i}{\hbar}[w, H] \tag{4.34}$$

(note the order of the operators in the commutator, in comparison with equation (4.18) for operators F).

The position and momentum operators $\mathbf{r}$ and $\mathbf{p}$ of a particle do not commute (equation (4.30)). It follows that they cannot be measured simultaneously. If we measure the position with infinite accuracy, the momentum is completely undefinable; and viceversa. That means that when we measure the position of a particle we disturb it, by transferring to it an unknown momentum; and when we measure the momentum of a particle, we disturb its position. A state with well defined momentum is the state of an electron, for instance, accelerated in a finite space region by an electric field and then let free; a light ray reflected off an electron (in an ideal microscope) can indicate the electron position. The impossibility of knowing simultaneously position and momentum, *i.e.* position and velocity, of a quantum particle, means that the quantum particles have not a trajectory; we cannot (it is impossible to) define a trajectory for quantum particles. This is a surprising and far-reaching consequence of the quantization, in striking disagreement with Newton's Mechanics.

[7] L. Landau, "Das Dampfungsproblem in der Wellenmechanik", Z. Phys. **45** 430 (1927).

Let $\Delta F = F - \overline{F}$ and $\Delta G = G - \overline{G}$ for two operators F and G. Let

$$((\Delta F - i\lambda \Delta G)\varphi, (\Delta F - i\lambda \Delta G)\varphi) \geq 0 \ , \tag{4.35}$$

where λ is a parameter; from this equation we get easily

$$(\delta F)^2 - i\lambda \overline{C} + \lambda^2 (\delta G)^2 \geq 0 \ , \tag{4.36}$$

where $C = [F, G]$ is the commutator of F with G; this equation implies

$$\delta F \cdot \delta G \geq \frac{1}{2} \left|\overline{C}\right| \ ; \tag{4.37}$$

for canonical conjugate variables $\mathbf{p}$ and $\mathbf{r}$ $C = -i\hbar$ and we have

$$\delta p \cdot \delta r \geq \frac{1}{2}\hbar \ , \ \delta p_x \cdot \delta x \geq \frac{1}{2}\hbar \ , \ etc \ . \tag{4.38}$$

These are Heisenberg's uncertainty (indeterminacy) relations.[8] In the quasi-classical case $\delta F \cdot \delta G \simeq \left|\overline{C}\right|/2$; in particular, if $G = H$ (the hamiltonian) and F does not depend on time, we have

$$\delta F \cdot \delta E \geq \frac{1}{2}\hbar \left|\overline{\dot{F}}\right| \ . \tag{4.39}$$

The mechanical motion proceeds with a fundamental indeterminacy, which is controlled by Planck's constant of mechanical action.

From equation (4.39) we may note that the most accurate measurement of F implies a certain time interval δt, such that $\delta E \cdot \delta t$ is of the order $\hbar$; we write

$$\delta E \cdot \delta t \simeq \hbar \ ; \tag{4.40}$$

since the minimum uncertainty in energy is $\delta E \simeq \delta v \cdot \delta p$, where v is the velocity, we have $\delta v \cdot \delta p \cdot \delta t \simeq \hbar$; since the speed of light in vacuum c is the highest possible velocity, we have $\delta p \cdot \delta t \simeq \hbar/c$, which imposes an important constraint on the measurement process of the momentum. Similarly, from $\delta p \cdot \delta x \simeq \hbar$ we get $\delta x \simeq \hbar/mc$ (which is called the Compton wavelength of the particle); and $\delta p \simeq mc$ (which is the ultrarelativistic case); as well as, $\delta E \cdot \delta x \simeq c\hbar$, and $\delta E \simeq mc^2$,

[8] W. Heisenberg, "Uber den anschaulichen Inhalt der quantentheoretischen Kinematik und Mechanik", Z. Phys. **43** 172 (1927).

$\delta t \simeq \hbar/mc^2$; for higher relativistic momenta, or energies, the Quantum Mechanics becomes Classical Mechanics. We can see that the theory of relativity combined with quantum mechanics exhibits fundamental constraints related to space, time and energy.[9]

4.4 The linear harmonic oscillator

The hamiltonian of the linear harmonic oscillator is

$$H = \frac{1}{2m}p^2 + \frac{1}{2}m\omega^2 x^2 \ , \tag{4.41}$$

where m is the mass, ω is the frequency, x is the coordinate (position) and p is the momentum; x, p and H are viewed as operators (matrices). The equations of motion are

$$\dot{x} = \frac{i}{\hbar}[H, x] = p/m \ , \ \dot{p} = \frac{i}{\hbar}[H, p] = -m\omega^2 x \ , \tag{4.42}$$

where we use the commutation relation $[p, x] = -i\hbar$; it follows

$$\ddot{x} + \omega^2 x = 0 \ , \tag{4.43}$$

or

$$(\ddot{x})_{nm} + \omega^2 x_{nm} = 0 \ ; \tag{4.44}$$

making use of

$$\dot{x}_{nm} = \frac{i}{\hbar}(E_n - E_m)x_{nm} = i\omega_{nm}x_{nm} \ , \tag{4.45}$$

where we denoted $(E_n - E_m)/\hbar$ by ω_{nm}, we get

$$(\omega_{nm}^2 - \omega^2)x_{nm} = 0 \ . \tag{4.46}$$

It follows $\omega_{nm} = \pm\omega$; we choose $\omega_{n+1,n} = \omega$ and $\omega_{n-1,n} = -\omega$; therefore, the matrix elements are

$$\dot{x}_{n+1,n} = i\omega x_{n+1,n} \ , \ \dot{x}_{n-1,n} = -i\omega x_{n-1,n} \ ; \tag{4.47}$$

[9] L. Landau and R. Peierls, "Erweiterung des Unbestimmtheitsprinzips fur die relativistische Quantentheorie", Z. Phys. **69** 56 (1931).

we note that $x_{nm} = x_{mn}$. Now we write the commutation relation $[p, x] = -i\hbar$ with $p = m\dot{x}$ in matricial form:

$$\begin{gathered}\dot{x}_{n,n-1}x_{n-1,n} + \dot{x}_{n,n+1}x_{n+1,n} - \\ -x_{n,n-1}\dot{x}_{n-1,n} - x_{n,n+1}\dot{x}_{n+1,n} = -\tfrac{i\hbar}{m} \ ,\end{gathered} \tag{4.48}$$

or

$$x^2_{n+1,n} - x^2_{n,n-1} = \frac{\hbar}{2m\omega} \ ; \tag{4.49}$$

hence,

$$x_{n,n-1} = \sqrt{\frac{\hbar n}{2m\omega}} \ ; \tag{4.50}$$

this equation gives all the other matrix elements. The energy levels are

$$E_n = H_{nn} = m\omega^2[(x_{n+1,n})^2 + (x_{n,n-1})^2] = \hbar\omega(n + 1/2) \ ; \tag{4.51}$$

except for the 1/2-term, this is precisely the Planck's proposal for the quantization of the oscillators energy; the 1/2-term does not change the Planck's energy-distribution law for the black body. It is worth noting that the ground state has a finite energy $\hbar\omega/2$; it is also called the vacuum energy.

4.5 Quantization of the angular momentum

The orbital angular momentum is

$$\mathbf{L} = \mathbf{r} \times \mathbf{p} \ , \ L_i = \varepsilon_{ijk}x_jp_k \ , \tag{4.52}$$

where ε_{ijk} is the antisymmetric tensor of rank three and i, j, k denote the cartesian components. Making use of $[p_i, x_j] = -i\hbar\delta_{ij}$, we get

$$[L_i, x_j] = i\hbar\varepsilon_{ijk}x_k \ , \ [L_i, p_j] = i\hbar\varepsilon_{ijk}p_k \tag{4.53}$$

and

$$[L_i, L_j] = i\hbar\varepsilon_{ijk}L_k \ ; \tag{4.54}$$

in addition

$$[L^2, L_i] = 0 \ , \tag{4.55}$$

where $L^2 = L_x^2 + L_y^2 + L_z^2$; equations (4.54) and (4.55) are preserved for intrinsic momenta.

It follows that L_i cannot be measured simultaneously; but L^2 and, say, L_z can. Let φ_m be the eigenvectors of L_z, corresponding to eigenvalues α_m,

$$L_z\varphi_m = \alpha_m\varphi_m \ ; \tag{4.56}$$

let us introduce the raising and lowering operators

$$L_\pm = L_x \pm iL_y \ , \tag{4.57}$$

which have the properties

$$[L_z, L_+] = \hbar L_+ \ , \ [L_z, L_-] = -\hbar L_- \ ; \tag{4.58}$$

then

$$L_z(L_+\varphi_m) = (\alpha_m + \hbar)L_+\varphi_m \tag{4.59}$$

and

$$L_z(L_-\varphi_m) = (\alpha_m - \hbar)L_-\varphi_m \ ; \tag{4.60}$$

it follows that we can construct the eigenvectors of the L_z by using the operators $L_\pm$ and the corresponding eigenvalues differ from one another by integral multiples of $\hbar$. On the other hand, since $L^2 - L_z^2 = L_x^2 + L_y^2 \geq 0$, it follows that there must be a maximum eigenvalue α and a minimum eigenvalue $-\alpha$; and 2α must be an integral multiple of $\hbar$; therefore α must be of the form $\alpha = \hbar l$, where l is an integer or a half-integer, and $\alpha_m = \hbar m$, where m takes the values$-l$, $-l+1, ...l-1$, l. There exist $2l+1$ states for the angular momentum. In addition, since $L_-L_+ = L^2 - L_z^2 - L_z$ and $L_+\varphi_l = 0$, we have

$$L^2\varphi_l = \hbar^2 l(l+1)\varphi_l \tag{4.61}$$

and

$$L^2\varphi_m = \hbar^2 l(l+1)\varphi_m \ . \tag{4.62}$$

We have also

$$\begin{aligned}(L_+\varphi_m, L_+\varphi_m) &= (\varphi_{m+1}, L_+\varphi_m) = (\varphi_m, L_-L_+\varphi_m) = \\ &= \hbar^2[l(l+1) - m(m+1)] = \hbar^2(l-m)(l+m+1) \ ,\end{aligned} \tag{4.63}$$

hence

$$(L_+)_{m+1,m} = \hbar\sqrt{(l-m)(l+m+1)} \ ,$$
$$(L_+)_{m-1,m} = \hbar\sqrt{(l+m)(l-m+1)} \ ; \tag{4.64}$$

hence we get the matrix elements of $L_{x,y}$.

The classical variable canonical conjugate to L_z is the angle φ, such that $\{L_z, \varphi\} = 1$; it follows that $[L_z, \varphi] = -i\hbar$, a relation similar with that between p_x and x (similar relations exist for any component of the angular momentum and the corresponding angular variable). In the classical limit $\hbar \to 0$ and l, $m \to \infty$, such that $\mathbf{L}$ be finite.

We shall see that for the orbital momentum we must limit ourselves to integers l ($l = 0, 1, 2, ...$); for intrinsic angular momentum, *i.e.* for spins $\mathbf{S}$, we may take $S = 0, 1, 1/2, ...$; the eigenvalues of S^2 are $\hbar^2 S(S+1)$ and the eigenvalues of S_z are $\hbar m$, where m takes the values $-S, -S+1, ...S-1, S$ ($2S+1$ values); the spin is a characteristic of the particle, *i.e.* S is fixed; consequently, in the classical limit $\hbar \to 0$ the spin $\mathbf{S}$ is vanishing.

We note that L^2 may be viewed as the square of the vector $\mathbf{L}$, with L_z its z-component; therefore, we may have an angle θ such that $\cos\theta = L_z/L$, or $\cos\theta = m/\sqrt{l(l+1)}$. We can be interested in the addition of two angular momenta, $\mathbf{L} = \mathbf{L}_1 + \mathbf{L}_2$ (or $\mathbf{J} = \mathbf{L} + \mathbf{S}$, where $\mathbf{L}$ is the orbital momentum, $\mathbf{S}$ is the spin and $\mathbf{J}$ is the total momentum); then we see that M corresponding to L goes from $|m_1 - m_2|$ to $m_1 + m_2$, where $m_{1,2}$ correspond to $L_{1,2}$; it follows that the state vectors of $\mathbf{L}$ (*i.e.* the eigenvectors of L^2 and L_z), which are products of state vectors of $\mathbf{L}_{1,2}$ (*i.e.* eigenvectors of $L_{1,2}^2$ and $L_{1,2z}$) form separate sets of state vectors corresponding to the angular momenta $L = |L_1 - L_2|, |L_1 - L_2| + 1, ...L_1 + L_2$, for each L-value being $2L+1$ vectors labelled by $M = -L, -L+1, ...L$.[10] Sums of operators corresponding to distinct "systems", like $\mathbf{L}_1 + \mathbf{L}_2$ are called direct (or tensorial) sums, while products of the state vectors of the terms of the sums, which form the state vectors of the sums, are called direct (or tensorial) products.

[10] The coefficients in the $\mid L_1 m_1 > \mid L_2 m_2 >$-expansion of the states $\mid LM >$ are called Clebsch-Gordan coefficients (A. R. Edmonds, *Angular Momentum in Quantum Mechanics*, Princeton University Press, Princeton (1957)).

4.6 Hydrogen atom

Let

$$H = \frac{1}{2m}p^2 - \frac{\alpha}{r} \quad , \tag{4.65}$$

be the hamiltonian of a particle with mass m in a Coulomb potential; it corresponds to the relative motion of the electron about the atomic nucleus, which consists of a proton, in the hydrogen atom; in that case $\alpha = e^2$, where e is the electron charge; similarly, it may correspond to hydrogen-like atoms, which consist of an electron and a nucleus with charge $-Ze$, where Z is the atomic number; in that case $\alpha = Ze^2$. The quantization incorporated in the commutation relations $[p_i, x_k] = -i\hbar\delta_{ik}$, where i, k denote cartesian components and x_i are the position coordinates, leads to stationary states for the motion governed by H with energy levels $E_n = -\frac{m\alpha^2}{2\hbar^2 n^2}$, where $n = 1, 2, 3, ...$[11]

First, it is useful to introduce reduced (dimensionless) variables $\mathbf{r} = (\hbar^2/m\alpha)\mathbf{R}$, $\mathbf{p} = (m\alpha/\hbar)\mathbf{P}$, $H = (m\alpha^2/\hbar^2)h$, such that $[P_i, X_j] = -i\delta_{ij}$ and

$$h = \frac{1}{2}P^2 - \frac{1}{R} \quad ; \tag{4.66}$$

the energy levels of h will be $-1/2n^2$. $\hbar^2/me^2 \simeq 0.53$Å is called the Bohr radius.

The angular momentum

$$\mathbf{L} = \mathbf{R} \times \mathbf{P} \tag{4.67}$$

is a constant of motion (as for any centrally-symmetric potential), *i.e.* $\dot{\mathbf{L}} = i[h, \mathbf{L}] = 0$ ($\mathbf{L}$ is conserved during the motion); it follows that L^2, L_z and h have common eigenvectors.

It can easily be checked that the Lenz vector

$$\mathbf{M} = \frac{1}{2}(\mathbf{P} \times \mathbf{L} - \mathbf{L} \times \mathbf{P}) - \frac{\mathbf{R}}{R} \tag{4.68}$$

is also conserved ($[h, \mathbf{M}] = 0$) . We note that in comparison with its classical counterpart, the Lenz vector has symmetrized here, in order to ensure its hermiticity. In addition,

$$[L_i, M_j] = i\varepsilon_{ijk}M_k \quad , \tag{4.69}$$

[11] W. Pauli, "Uber das Wasserstoffspektrum vom Standpunkt der neuen Quantemechanik", Z. Phys. **36** 336 (1926).

as for any vector, according to equation (4.53); but $[L_i, M^2] = 0$, as for any scalar. It follows that h, L^2, M^2 have common eigenvectors (L_z and M_z do not contribute to eigenvalues, since they are not scalars). Making use of the definition of the Lenz vector we get

$$M^2 = 2H(L^2 + 1) + 1 \ , \tag{4.70}$$

a relation which differs from its classical counterpart by $L^2 + 1$ instead of L^2 (in the classical limit $L \to \infty$ and the two expressions coincide); we have also

$$[M_i, M_j] = -2i\varepsilon_{ijk} H L_k \ . \tag{4.71}$$

For the common eigenstates we may replace H by its eigenvalue $-E$ (with $E > 0$) and define

$$\mathbf{N} = \mathbf{M}/\sqrt{2E} \ , \tag{4.72}$$

such that

$$[N_i, N_j] = i\varepsilon_{ijk} L_k \ ; \tag{4.73}$$

from equation (4.70) we get

$$E = \frac{1}{2(L^2 + N^2 + 1)} \ . \tag{4.74}$$

Now we introduce two new operators

$$\mathbf{F} = \frac{1}{2}(\mathbf{L} + \mathbf{N}) \ , \ \ \mathbf{G} = \frac{1}{2}(\mathbf{L} - \mathbf{N}) \ , \tag{4.75}$$

such that $L^2 + N^2 = 2(F^2 + G^2)$; it is easy to check that $\mathbf{F}$ and $\mathbf{G}$ are two independent angular momenta, according to their commutation relations, and $F^2 = G^2$ (since $\mathbf{LN} + \mathbf{NL} = 0$). It follows that $F^2 = G^2 = f(f + 1)$, $f = 0,\ 1/2,\ 1, ...$ and $L^2 + N^2 = 4f(f + 1)$; the energy levels given by equation (4.74) become

$$E = \frac{1}{2(2f + 1)^2} = \frac{1}{2n^2} \ , \tag{4.76}$$

where $n = 2f + 1 = 0,\ 1,\ 2, ...$ n is called the principal quantum number; the angular quantum number l goes from zero to $2f = n - 1$ (from $\mathbf{L} = \mathbf{F} + \mathbf{G}$ and $F^2 = G^2$). We can see that the energy levels do not depend on m, *i.e.* they are degenerate with respect to the orientation of the angular momentum, as expected for a centrally-symmetric potential; but they do not depend on l either, they are degenerate with respect to the angular momentum, which is a property of the Coulomb potential (expressed by the Lenz vector).

4.7 Wave function. Schrodinger's wave mechanics

In equation (4.26) we associate the differential operator $i\hbar\frac{\partial}{\partial t}$ to the hamiltonian H,

$$H \to i\hbar\frac{\partial}{\partial t} \quad , \tag{4.77}$$

and write the Schrodinger equation

$$i\hbar\frac{\partial}{\partial t}\psi = H\psi \quad ; \tag{4.78}$$

it follows that we consider the state vectors as being functions of the time t. The fundamental canonical quantum-mechanical commutation relation $[\mathbf{p}, \mathbf{r}] = -i\hbar$ (viewed as $[p_i, x_j] = -i\hbar\delta_{ij}$, with usual notations) sugests a further development; we may imagine that the position operator $\mathbf{r}$ has a continuous set of eigenvalues, denoted by the same symbol $\mathbf{r}$, which this time is a c-number; in the canonical commutation relation of $\mathbf{p}$ with $\mathbf{r}$ we may consider $\mathbf{r}$ as a c-number; then, we have the association

$$\mathbf{p} \to -i\hbar\frac{\partial}{\partial \mathbf{r}} \tag{4.79}$$

(or $\mathbf{p} \to -i\hbar grad$); it follows that we are led to consider the state vectors as functions $\psi(t, \mathbf{r})$ of the time t and position $\mathbf{r}$; then, the hamiltonian of a particle with mass m in a potential $V(\mathbf{r})$ becomes

$$H = -\frac{\hbar^2}{2m}\Delta + V(\mathbf{r}) \quad , \tag{4.80}$$

and the Schrodinger equation becomes

$$i\hbar\frac{\partial}{\partial t}\psi = [-\frac{\hbar^2}{2m}\Delta + V(\mathbf{r})]\psi \quad ; \tag{4.81}$$

if ψ is an eigenvector of H, corresponding to energy E, then $\psi = e^{-\frac{i}{\hbar}Et}\varphi$ and

$$[-\frac{\hbar^2}{2m}\Delta + V(\mathbf{r})]\varphi = E\varphi \ ; \qquad (4.82)$$

equation (4.82) is called (time-independent) Schrodinger equation, while equations (4.78) and (4.81) may be called time-dependent Schrodinger equations.[12] The state vectors $\psi(t,\mathbf{r})$ are called wavefunctions. It follows that we associate waves $\psi(t,\mathbf{r})$ to material particles, as de Broglie hypothesized. We can see that the Schrodinger equation is an equation for eigenfunctions φ and eigenvalues E; its spectrum may be both continuous and discrete. The wavefunctions φ corresponding to well-defined (constant) energies describe the stationary states. The eigenfunctions of $\mathbf{r}$ corresponding to the eigenvalue $\mathbf{r}_0$ are $\delta(\mathbf{r}-\mathbf{r}_0)$.

With Dirac's notations we may write $\psi(t,\mathbf{r}) =< \mathbf{r} \mid \psi(t) >$, and we may see that $\psi(t,\mathbf{r})$ is the probability amplitude to get a particle which is in state ψ at position $\mathbf{r}$ (at time t) (or to get at $\mathbf{r}$ the particle which is in state ψ); $|\psi(t,\mathbf{r})|^2$ is the probability to get at $\mathbf{r}$ the particle which is in state ψ (at time t). It is admitted that this interpretation has been given by Born.[13] Integrals of the form

$$\int d\mathbf{r}\psi^*(t,\mathbf{r})\varphi(t,\mathbf{r}) \qquad (4.83)$$

can be viewed as scalar products; the closure relation requires

$$\int d\mathbf{r}\,|\psi(t,\mathbf{r})|^2 = 1 \ , \qquad (4.84)$$

which is called the normalization condition; the orthonormality rela-

[12] E. Schrodinger, "Quantisierung als Eigenwertproblem (I)", Ann. Physik **79** (4) 361 (1926); "Quantisierung als Eigenwertproblem (II)", Ann. Physik **79** (4) 489 (1926); "Uber das Verhaltnis der Heisenberg-Born-Jordanschen Quantenmechanik zu der meinen", Ann. Physik **79** (4) 734 (1926); "Quantisierung als Eigenwertproblem (III)", Ann. Physik **80** (4) 437 (1926); "Quantisierung als Eigenwertproblem (IV)", Ann. Physik **81** (4) 109 (1926).

[13] M. Born, "Zur Quantenmechanik der Stossvorgange", Z. Phys. **37** 863 (1926).

tions can be written as

$$\begin{gathered}\int d\mathbf{r}\psi_E^*(t,\mathbf{r})\psi_{E'}(t,\mathbf{r}) = e^{\frac{i}{\hbar}((E-E^{'})t}\int d\mathbf{r}\varphi_E^*(\mathbf{r})\varphi_{E'}(\mathbf{r}) = \\ = e^{\frac{i}{\hbar}((E-E^{'})t}\delta(E-E^{'}) = \delta(E-E^{'}) \ .\end{gathered} \tag{4.85}$$

If the energies are discrete and imply sets of integers n, then the normalization of the wavefunctions may be made to these integers. The matrix elements of the operators are written as

$$\int d\mathbf{r}\psi^*(t,\mathbf{r})F(t,\mathbf{r})\varphi(t,\mathbf{r}) \ , \tag{4.86}$$

or, more generally,

$$\int d\mathbf{r}d\mathbf{r}^{'}\psi^*(t,\mathbf{r})F(t,\mathbf{r},\mathbf{r}^{'})\varphi(t,\mathbf{r}) \ , \tag{4.87}$$

which are bilinear forms of wave functions; if the wavefunctions are labelled by some sets of integers n, m, then we get matrix elements of the form F_{nm}; t in $F(t,\mathbf{r},\mathbf{r}^{'})$ denotes the intrinsic time dependence of the operator (if any), not generated by the quantum-mechanical dynamics. Matrix elements of the form F_{nm} can be written as

$$F_{nm} = \int d\mathbf{r}d\mathbf{r}^{'} < n \mid \mathbf{r} >< \mathbf{r} \mid F(t) \mid \mathbf{r}^{'} >< \mathbf{r}^{'} \mid m > \ , \tag{4.88}$$

whence the matrix elements $< \mathbf{r} \mid F(t) \mid \mathbf{r}^{'} >= F(t,\mathbf{r},\mathbf{r}^{'})$. We can see that the wavefunction is determined up to a phase factor. The mean values of the operators are of the form

$$\overline{F} = \int d\mathbf{r}\psi^*(t,\mathbf{r})F(t,\mathbf{r},\mathbf{r}^{'})\psi(t,\mathbf{r}^{'}) \mid_{\mathbf{r}=\mathbf{r}'} \ . \tag{4.89}$$

We note that equally well we may consider the continuous set of eigenvalues $\mathbf{p}$ for the momentum operator $\mathbf{p}$; in that case the commutation relation would imply $\mathbf{r} \rightarrow i\hbar\frac{\partial}{\partial\mathbf{p}}$; this is called the momentum representation (or $\mathbf{p}$-representation), in contrast with the former representation $\mathbf{p} \rightarrow i\hbar\frac{\partial}{\partial\mathbf{r}}$ which is called $\mathbf{r}$-representation, or position representation (or q-representation, since q is reminiscent of the coordinate notation in Classical Mechanics). Also, we note that the time

t is a parameter in Quantum Mechanics, such that a "representation" of the form $t \rightarrow -i\hbar\frac{\partial}{\partial E}$ has only a limited validity and usefulness.

We note here a point related to the adjointness of the differential operators; we have

$$\int d\mathbf{r}\psi^*(-i\hbar\tfrac{\partial}{\partial \mathbf{r}})\varphi = -i\hbar\psi^*\varphi\mid_S + \int d\mathbf{r}(i\hbar\tfrac{\partial}{\partial \mathbf{r}})\psi^*\varphi =$$
$$= -i\hbar\psi^*\varphi\mid_S + \int d\mathbf{r}(-i\hbar\tfrac{\partial}{\partial \mathbf{r}}\psi)^*\varphi \ , \tag{4.90}$$

where S denotes the boundary of the integration domain; the integration by parts gives the transposed operator, while the adjointness is ensured by taking the complex conjugation too. The self-adjointness (hermiticity) is ensured by zero contribution of the boundary; it follows that the wavefunctions must be vanishing at the boundary (at infinity, for instance), or be periodical over the integration domain (we must impose in that case the so-called cyclic boundary conditions). In addition, the wavefunctions, as solutions of the Schrodinger equation, which is a second-order differential equation, must be continuous and have continuous derivatives (they must be differentiable); special cases of discontinuous potentials require special considerations. Since the Schrodinger equation is linear, its solutions are superpositions of wavefunctions. Of course, the wavefunctions must be single-valued.

Let us multiply Schrodinger equation (4.81) by ψ^* on the left and the the Schrodinger equation for ψ^* by ψ on the right, and subtract the results; we get

$$\psi^* \mid \ i\hbar\tfrac{\partial\psi}{\partial t} = -\tfrac{\hbar^2}{2m}\Delta\psi + V\psi \ ,$$
$$-i\hbar\tfrac{\partial\psi^*}{\partial t} = -\tfrac{\hbar^2}{2m}\Delta\psi^* + V\psi^* \ \mid \psi \tag{4.91}$$

and

$$\frac{\partial}{\partial t}\left|\psi\right|^2 + \frac{i\hbar}{2m}div\left[grad\psi^* \cdot \psi - \psi^* grad\psi\right] \ ; \tag{4.92}$$

we introduce the notations

$$\rho = \left|\psi\right|^2 \tag{4.93}$$

for the probability density and

$$\mathbf{j} = \frac{i\hbar}{2m}\left[grad\psi^* \cdot \psi - \psi^* grad\psi\right] \tag{4.94}$$

for the particle flow, *i.e.* the number of particles per unit time per unit area perpendicular to $\mathbf{r}$; equation (4.92) is the continuity equation

$$\frac{\partial \rho}{\partial} + div\mathbf{j} = 0 \quad , \tag{4.95}$$

which ensures the conservation of the number of particles; ρ may be viewed as the density of particles (at $\mathbf{r}$ and t), *i.e.* the number of particles per unit volume (or the amount of a particle per unit volume); note that the normalization condition given by equation (4.84) tells that $\psi \sim 1/\sqrt{V}$, where V is the volume of the integration domain (*i.e.*, the domain where the motion exists); the quantum-mechanical wavefunction provides a fluid picture for the quantum particles. The continuity and derivability of the wavefunction ensure the physical condition of the conservation of matter.

Let us take the mean value of the Schrodinger equation (4.82); we get

$$\begin{gathered}(\psi, [-\tfrac{\hbar^2}{2m}\Delta + V(\mathbf{r}) - E]\varphi) = \\ \int d\mathbf{r}\left[\tfrac{\hbar^2}{2m}\left|grad\psi\right|^2 + V\left|\psi\right|^2 - E\left|\psi\right|^2\right] \quad ;\end{gathered} \tag{4.96}$$

we can see that the Schrodinger equation follows from the requirement that the mean energy $p^2/2m + V$, with $p = \hbar\left|grad\psi\right|$ has an extremum for normalized wavefunctions $\left|\psi\right|^2 = const$; this was Schrodinger's original derivation of his equation; with trial wavefunctions the method can be used to get the ground-state; excited states can be get by using trial wavefunctions orthogonal to the ground-state (this is sometimes called the Rayleigh-Ritz principle). Of course, the energy eigenvalues are greater than the minimum of the potential. The continuous energy spectrum is positive (since the particles are free at infinity), while the bound states have discrete negative energies. The continuity of the wavefunctions implies that quantum particles may be in regions classically unavailable, where the potential energy is greater than their energy; the momentum is then a purely imaginary quantity.

We can see that the ground-state wavefunction (for one particle) has no nodes (no zeros); and it cannot be degenerate; the excited states have wavefunctions with nodes (this can be seen from the kinetic-energy operator $\sim \left|grad\psi\right|^2$).

4 Quantum Mechanics

Let us consider Newton's classical equation of motion of a particle with mass under the action of a force $\mathbf{F}$: $d\mathbf{p}/dt = \mathbf{F}$, where $\mathbf{p}$ is the momentum of the particle; let us consider that the particle is distributed in space, like a fluid; in addition let us assume that there exists a potential function V, such that $\mathbf{F} = -gradV$; this V may depend on the time, in general; then, its equation of motion becomes

$$\left(\frac{\partial}{\partial t} + \mathbf{v}grad\right)\mathbf{p} = -gradV \quad , \tag{4.97}$$

where $\mathbf{v}$ is the transport velocity; let us multiply this equation by $i\hbar$, to get

$$\left(i\hbar\frac{\partial}{\partial t} + i\hbar\mathbf{v}grad\right)\mathbf{p} = -i\hbar gradV \quad ; \tag{4.98}$$

let us view the momentum $\mathbf{p}$ in this equation the quantum-mechanical operator $-i\hbar grad$; then, we may write the above equation as

$$-i\hbar grad\left(i\hbar\frac{\partial}{\partial t} + i\hbar\mathbf{v}grad\right) = -i\hbar gradV \quad , \tag{4.99}$$

or

$$i\hbar\frac{\partial}{\partial t} + i\hbar\mathbf{v}grad = V \quad ; \tag{4.100}$$

it looks like a wave equation, where $\mathbf{v}$ is the phase velocity; for small (or constant) V we may write $\mathbf{v} = \mathbf{p}/2m$, such that the above equation becomes the Schrodinger equation

$$i\hbar\frac{\partial}{\partial t} = -\frac{\hbar^2}{2m}\Delta + V \quad ; \tag{4.101}$$

it exhibits, in fact, the equation of the energy (which may not be conserved); the time-dependent function which may appear on pasing from equation (4.99) to equation (4.100) may be set equal to zero. The procedure described above may also be read in the opposite direction, leading from the Schrodinger equation to the classical Newton's equation of motion. It is worth noting that V should not vary too fast or too abruptly.

4.8 Free particle

We can check immediately that the eigenfunctions of the momentum are

$$\varphi_{\mathbf{p}}(\mathbf{r}) = \frac{1}{\sqrt{V}} e^{\frac{i}{\hbar}\mathbf{pr}} \quad , \tag{4.102}$$

where the constant $1/\sqrt{V}$ is arbitrary; the orthogonality condition is

$$\int d\mathbf{r}\varphi^*_{\mathbf{p}'}(\mathbf{r})\varphi_{\mathbf{p}}(\mathbf{r}) = \frac{1}{V}\int d\mathbf{r}e^{\frac{i}{\hbar}(\mathbf{p}-\mathbf{p}')\mathbf{r}} = \frac{(2\pi)^3\hbar^3}{V}\delta(\mathbf{p}-\mathbf{p}') \ ; \tag{4.103}$$

the plane waves given by equation (4.102) may be viewed as being enclosed in a parallelepipedic volume V with sides L_i; the cyclic boundary conditions require

$$p_i = \hbar k_i \ , \ k_i = \frac{2\pi}{L_i} n_i \quad , \tag{4.104}$$

where $\mathbf{k}$ is a wavevector and n_i are integers; n_i must be non-zero, because a particle at rest (with momentum zero) does not exist in Quantum Mechanics (due to the uncertainty relations); and motion confined strictly to a plane or a line is unphysical; it may lead to meaningless results, if special conditions are not added. Making use of equations (4.104), the orthonormality conditions given by equation (4.103) become

$$\int d\mathbf{r}\varphi^*_{\mathbf{p}'}(\mathbf{r})\varphi_{\mathbf{p}}(\mathbf{r}) = \delta_{\mathbf{n},\mathbf{n}'} \quad , \tag{4.105}$$

where $\delta_{\mathbf{n},\mathbf{n}'}$ is the Kronecker symbol with sets of numbers $\mathbf{n} = (n_i)$, $i = 1, 2, 3$; we note that the infinitesimal number of states is $d\mathbf{n} = \frac{V}{(2\pi)^3}d\mathbf{k} = \frac{V}{(2\pi)^3\hbar^3}d\mathbf{p}$. From equations (4.104) we see that the particle is a three-dimensional wave with wavelengths $\lambda_i = L_i/n_i$ and the momentum is $p_i = h/\lambda_i$, in agreement with de Broglie's hypothesis. The hamiltonian of a free particle is

$$H = \frac{1}{2m}p^2 = -\frac{\hbar^2}{2m}\Delta \ ; \tag{4.106}$$

we can see that the momentum eigenfunctions are also energy eigenfunctions with the eigenenergies

$$E_{\mathbf{p}} = \frac{1}{2m}p^2 = \frac{\hbar^2 k^2}{2m} = \frac{\hbar^2(2\pi)^2}{2m}\left[\frac{n_1^2}{L_1^2} + \frac{n_2^2}{L_2^2} + \frac{n_3^2}{L_3^2}\right] \ ; \tag{4.107}$$

the time-dependent eigenfunctions are

$$\psi_{\mathbf{p}}(t,\mathbf{r}) = \frac{1}{\sqrt{V}} e^{-\frac{i}{\hbar}E_{\mathbf{p}}t+\frac{i}{\hbar}\mathbf{pr}} = \frac{1}{\sqrt{V}} e^{-\frac{i}{\hbar}\frac{p^2}{2m}t+\frac{i}{\hbar}\mathbf{pr}} \ ; \tag{4.108}$$

we note the (classical) mechanical action $S = -Et + \mathbf{pr}$ of a free particle in the exponent of the wavefunction. We note that $e^{\frac{i}{\hbar}\mathbf{pr}} = e^{i\mathbf{kr}}$ is a travelling wave which propagates along the direction of $\mathbf{p}$ ($\mathbf{k}$) (direction along which the phase $\mathbf{kr}$ increases); similarly, the complex conjugate $e^{-\frac{i}{\hbar}\mathbf{pr}} = e^{-i\mathbf{kr}}$ propagates in the opposite direction. By convention, $e^{-\frac{i}{\hbar}Et} = e^{-i\omega t}$, where we denoted $E = \hbar\omega > 0$, is a wave which propagates in the future; and $e^{\frac{i}{\hbar}Et} = e^{i\omega t}$ is a wave which propagates in the past; these waves are delocalized both in space and time. In Fourier expansions of the type

$$f(t,\mathbf{r}) = \frac{1}{(2\pi)^4}\int d\omega d\mathbf{k} f(\omega,\mathbf{k}) e^{-i\omega t+i\mathbf{kr}} \tag{4.109}$$

for an arbitrary function f (Fourier decomposable), we include plane waves which propagate in all directions, both in the future and in the past.

The plane waves are degenerate with respect to the direction of $\mathbf{p}$. Let us consider wavefunctions given by

$$\varphi_p = \frac{C}{4\pi}\int do e^{\frac{i}{\hbar}\mathbf{pr}} = C\frac{\sin kr}{kr} \ , \tag{4.110}$$

where C is a constant of normalization and o is the solid angle between $\mathbf{k}$ and $\mathbf{r}$; making use of the laplacian in spherical coordinates, we get the hamiltonian of the free particle

$$\begin{gathered} H = -\frac{\hbar^2}{2m}\Delta = -\frac{\hbar^2}{2m}\left[\frac{1}{r^2}\frac{\partial}{\partial r}(r^2\frac{\partial}{\partial r})+\right. \\ \left.+\frac{1}{r^2\sin\theta}\frac{\partial}{\partial\theta}(\sin\theta\frac{\partial}{\partial\theta}) + \frac{1}{r^2\sin^2\theta}\frac{\partial^2}{\partial\varphi^2}\right] \ ; \end{gathered} \tag{4.111}$$

we can check easily that φ_p are eigenfunctions of this hamiltonian,

$$H\varphi_p = \frac{\hbar^2 k^2}{2m}\varphi_p \tag{4.112}$$

(the angles θ and φ do not enter the radial wavefunctions φ_p); it is easy to check that

$$\int d\mathbf{r}\varphi_p^*\varphi_{p'} = \frac{2\pi^2C^2}{k^2}\delta(k-k^{'}) = \frac{2\pi^2C^2\hbar^2}{mk}\delta(E-E^{'}) \ ; \qquad (4.113)$$

we may choose the constant C such that the wavefunctions be normalized to energy, and we may denote them φ_E. It is worth noting that the wavefunctions given by equation (4.110) are superpositions of two spherical waves, an outgoing one e^{ikr}/r and an incoming one e^{-ikr}/r.

From equation (4.108) we can see that the free particle current density is

$$\mathbf{j} = \frac{\mathbf{p}}{m}\left|\psi\right|^2 = \mathbf{v}\left|\psi\right|^2 \quad , \qquad (4.114)$$

where $v = \partial\mathbf{E}/\partial\mathbf{p} = \partial\omega/\partial\mathbf{k}$ is the group velocity (distinct from the phase velocity $\mathbf{p}/2m$).

4.9 Wave packets

The wavefunction

$$\psi_{\mathbf{p}}(t,\mathbf{r}) = \frac{1}{\sqrt{V}}e^{-\frac{i}{\hbar}E_{\mathbf{p}}t+\frac{i}{\hbar}\mathbf{pr}} = \frac{1}{\sqrt{V}}e^{-i\omega(k)t+i\mathbf{kr}} \ ; \qquad (4.115)$$

for a free particle, with $\mathbf{p} = \hbar\mathbf{k}$ and $E_{\mathbf{p}} = \hbar\omega(k) = \hbar^2k^2/2m$, is a function extended in the whole space. Let us consider a free particle localized in a small volume $\Delta\mathbf{r}$; its wavefunction can be represented as a Fourier expansion

$$\psi(t,\mathbf{r}) = \frac{1}{(2\pi)^{3/2}}\int d\mathbf{k}A(\mathbf{k})e^{-i\omega(k)t+i\mathbf{kr}} \quad , \qquad (4.116)$$

with the normalization condition

$$\int d\mathbf{r}\left|\psi\right|^2 = \int d\mathbf{k}\left|A(\mathbf{k})\right|^2 = 1 \ ; \qquad (4.117)$$

such a wavefunction is called a wavepacket. If the wavefunction is localized over the small volume $\Delta\mathbf{r}$, the amplitudes $A(\mathbf{k})$ should be vanishing for large deviations of $\mathbf{k}$ from an arbitrary $\mathbf{k}_0$; otherwise, the rapid oscillations would destroy the wavefunction; it follows that $\mathbf{q} = \mathbf{k} - \mathbf{k}_0$ should extend up to values of the order $\Delta k_i \simeq 1/\Delta x_i$, where i denotes the cartesian coordinates and x_i are the position coordinates; and $A(\mathbf{k})$ should vary slowly inside this range; we can see that we get the uncertainty relations $\Delta p_i \Delta x_i \simeq \hbar$; we have restricted as much as possible the $\mathbf{k}$-extension; such that, in fact, we have $\Delta p_i \Delta x_i \geq \hbar$. If we consider the inequality

$$\int dx \left| \lambda x \psi + \frac{d\psi}{dx} \right|^2 \geq 0 \tag{4.118}$$

with $(\delta x)^2 = \int dx \cdot x^2 |\psi|^2$, we get $\delta x \cdot \delta p_x \geq \hbar/2$. Similar uncertainty relations hold for energy and time, by considering temporal wavepackets of the form $\int d\omega e^{-i\omega t}$. If we try to confine a quantum particle to a small spatial region with dimension Δx and the particle is described by a wavefunction, then the wavelength must be at most $\lambda \simeq 2\Delta x$, which gives a momentum $p \simeq h/2\Delta x$ at least; of course, we may also have wavelengths of the order $\lambda_n \simeq 2\Delta x/n$, where n is any positive integer (Δx must be a multiple of half-wavelengths), which correspond to momenta $p_n \simeq hn/2\Delta x$. This uncertainty in momentum induces an uncertainty in energy, through $\Delta E = v\Delta p = \Delta x \cdot \Delta p/\Delta t$, whence $\Delta E \cdot \Delta t > h/2$. It is worth noting that a perturbation which lasts a short time Δt induces a large uncertainty $\Delta E \simeq h/\Delta t$, irrespective of its strength.

Let us write the expansion

$$\begin{aligned} -\omega(k)t + \mathbf{kr} &= -\omega(|\mathbf{k}_0 + \mathbf{q}|)t + (\mathbf{k}_0 + \mathbf{q})\mathbf{r} = \\ &= -\omega_0 t + \mathbf{k}_0\mathbf{r} + (\mathbf{r} - \mathbf{v}_0 t)\mathbf{q} - \tfrac{\hbar q^2 t}{2m} \ , \end{aligned} \tag{4.119}$$

where

$$\omega_0 = \omega(k_0) \ , \ \mathbf{v}_0 = \frac{\partial \omega_0}{\partial \mathbf{k}} = \frac{\hbar \mathbf{k}_0}{m} \ ; \tag{4.120}$$

the wavefunction given by equation (4.116) becomes

$$\psi(t, \mathbf{r}) = \frac{e^{-i\omega_0 t + i\mathbf{k}_0\mathbf{r}}}{(2\pi)^{3/2}} \int d\mathbf{q} A(\mathbf{k}_0 + \mathbf{q}) e^{i(\mathbf{r} - \mathbf{v}_0 t)\mathbf{q} - i\frac{\hbar q^2 t}{2m}} \ ; \tag{4.121}$$

for a very narrow localization we may extend the **q**-integration to infinity and neglect the **q**-dependence in $A(\mathbf{k}_0+\mathbf{q})$; for short times, we may also neglect the q^2-term in the exponent, and get

$$\psi(t,\mathbf{r}) = (2\pi)^{3/2}A(\mathbf{k}_0)e^{-i\omega_0 t+i\mathbf{k}_0\mathbf{r}}\delta(\mathbf{r}-\mathbf{v}_0 t) \ , \tag{4.122}$$

which is the wavefunction of a free particle localized at $\mathbf{r}=\mathbf{v}_0 t$, *i.e.* a "classical" particle moving with velocity $\mathbf{v}_0$; noteworthy, $\mathbf{v}_0$ is the group velocity of the waves. For longer times we include the contribution of the q^2-term in the exponent and get from equation (4.121)

$$\psi(t,\mathbf{r}) = A(\mathbf{k}_0)e^{-i\omega_0 t+i\mathbf{k}_0\mathbf{r}}(m/i\hbar t)^{3/2}e^{i\frac{m(\mathbf{r}-\mathbf{v}_0 t)^2}{2\hbar t}} \ ; \tag{4.123}$$

we can see that the wavepacket flattens in time, with slower oscillations; a localized quantum particle set free diffuses in the whole space.[14]

4.10 Classical limit

Any wavefunction ψ may be written as

$$\psi(t,\mathbf{r}) = A(t,\mathbf{r})e^{\frac{i}{\hbar}S(t,\mathbf{r})} \ , \tag{4.124}$$

where A is an amplitude and S is a phase with dimensions of mechanical action; inserted in the Schrodinger equation

$$i\hbar\frac{\partial\psi}{\partial t} = -\frac{\hbar^2}{2m}\Delta\psi + V\psi \tag{4.125}$$

and taking the real and imaginary parts we get easily

$$\begin{gathered}\frac{\partial A^2}{\partial t} + \frac{1}{m}div(A^2 gradS) = 0 \ , \\ \dot{S} + \frac{1}{2m}(gradS)^2 + V - \frac{\hbar^2}{2mA}\Delta A = 0 \ ;\end{gathered} \tag{4.126}$$

if we neglect the $\hbar^2$-contribution, we can see that S is the mechanical action with energy $E=-\dot{S}=-\partial S/\partial t$ and momentum $\mathbf{p}=gradS$;

[14] It is easy to see that the technique described here can also be used for viewing superpositions of waves as (quasi-) particles.

the first equation (4.126) is the continuity equation for $|\psi|^2 = A^2$, which moves with (the group) velocity $\mathbf{v} = \mathbf{p}/m$, while the second equation (4.126) (without the $\hbar^2$-term) is the Hamilton-Jacobi equation. It follows that the classical limit is obtained for $\hbar \rightarrow 0$, where the phase $\phi = S/\hbar$ is infinite and the wavefunction oscillates very rapidly; equally well, the classical limit is also attained for bodies with a large mass, with the amplitude of their wavefunctions varying slowly in space and with the mechanical action S varying greatly in space; we recognize in this latter case the macroscopic bodies, where the amplitude is a measure of their "presence". We can see that in the classical (macroscopic) limit the Quantum Mechanics becomes identical with Newton's Classical Mechanics; it brings its specific quantum-mechanical features in the microscopic limit, where amounts of mechanical action of the order $\hbar$ matter. Noteworthy, the quantum-mechanical corrections to the mechanical action are of the order $\hbar^2$; and the quantum-mechanical motion is performed by the phase S of the wavefunction (the $\hbar^2$-term included).

Let us introduce the probability density $n = |\psi|^2 = A^2$ and the (group) velocity $\mathbf{v} = \mathbf{p}/m = \frac{1}{m} gradS$; the first equation (4.126) becomes the continuity equation of a fluid,

$$\frac{\partial n}{\partial t} + div(\mathbf{v}n) = 0 \quad , \tag{4.127}$$

or

$$\frac{\partial \rho}{\partial t} + div(\mathbf{v}\rho) = 0 \quad , \tag{4.128}$$

if we multiply by m and use the mass density $\rho = mn$. In the second equation (4.126) we take the *grad* and divide by m; we get

$$\frac{\partial \mathbf{v}}{\partial t} + (\mathbf{v}grad)\mathbf{v} = \frac{1}{m}\mathbf{F} + \frac{\hbar^2}{2m^2} grad \frac{\Delta n^{1/2}}{n^{1/2}} \quad , \tag{4.129}$$

where $\mathbf{F}$ is the force acting upon the particle; if we view the particle as a small, extended object, then we may view the force as arising from a pressure p; then, we may write

$$\mathbf{F} = -\int p d\mathbf{S} = -\int d\mathbf{r} gradp = -\Delta\mathcal{V} \cdot gradp = -\frac{1}{n} gradp \tag{4.130}$$

for a small volume $\Delta\mathcal{V}$ of the particle, surrounded by a smal surface with the surface element $d\mathbf{S}$ (where $p = nV$). As long as we view these small volumes (actually sufficiently large, as for a classical fluid) as basic entities, the $\hbar^2$-term in equation (4.129) becomes very small; we have a quasi-classical fluid, with quantum-mechanical corrections; its equation of motion

$$\frac{\partial \mathbf{v}}{\partial t} + (\mathbf{v} grad)\mathbf{v} = -\frac{1}{\rho} grad p + \frac{\hbar^2}{2m^2} grad \frac{\Delta n^{1/2}}{n^{1/2}} \tag{4.131}$$

is Euler's equation for the motion of a fluid, with a (small) quantum-mechanical correction. We can see that Schrodinger's equation may become Euler's equation in the quasi-classical limit, providing we view the particle as a fluid.

5 Schrodinger Equation

5.1 General properties

We discuss here the time-dependent Schrodinger equation

$$i\hbar\frac{\partial\psi}{\partial t} = H\psi \tag{5.1}$$

and the Schrodinger equation

$$H\psi = E\psi \tag{5.2}$$

for stationary states with a definite energy E ($\psi \sim e^{-\frac{i}{\hbar}Et}$), where

$$H = \frac{p^2}{2m} + V(\mathbf{r}) \tag{5.3}$$

is the hamiltonian of a particle with mass m and potential energy $V(\mathbf{r})$; t denotes the time, $\mathbf{r}$ is position, $\psi(t,\mathbf{r})$ is the wavefunction, $\mathbf{p} = -i\hbar\frac{\partial}{\partial\mathbf{r}}$ is the momentum ($p^2 = -\hbar^2\Delta$) and $\hbar \simeq 10^{-27}erg \cdot s$ is Planck's constant; equation (5.2) reads

$$-\frac{\hbar^2}{2m}\Delta\psi + V\psi = E\psi \ ; \tag{5.4}$$

it is an equation for the eigenvalues E (energy levels) and eigenfunctions ψ (two different energies correspond to orthogonal wavefunctions). Obviously, since $\hbar$ is very small, it is an equation for small (quantum) particles, moving in small potentials with low energies; also, it is obvious from

$$E = \frac{\hbar^2}{2m}\int d\mathbf{r} \mid grad\psi \mid^2 + \int d\mathbf{r}V \mid \psi \mid^2 \tag{5.5}$$

that E should be higher than the minimum value of V (the minimum value of E is the ground-state energy).

We note that V does not depend on time. Also, equation (5.4) is real, then ψ may be taken real (or real combinations for degenerate energy levels); equation (5.1) is symmetric for $t \to -t$, $\psi \to \psi^*$ (time-reversal invariance).

We note that ψ is defined over the whole space; since the classical p^2 is replaced by $-\hbar^2 \Delta\psi$, the quantum particle may enter classically inaccessible regions; then, since $\mid \psi \mid^2$ is a probability density, it follows that ψ must be single-valued, continuous and with continuous derivatives (for the unphysical case of an infinite potential barrier, ψ vanishes at the boundaries and its derivatives are discontinuous); it is not necessary for ψ to be finite in some points. The potentials go to zero at infinity, so, there, the wavefunction may go like $e^{\frac{i}{\hbar}\mathbf{pr}}$ as for a free particle (plane wave); then, the energy is $E = p^2/2m$, with any $\mathbf{p}$; the spectrum is positive, continuous and degenerate (infinite degeneration). We may impose a vanishing ψ at infinity; then, we have a bound state, the wavefunctions cannot be continuously and infinitesimally deformed, so the spectrum is discrete and can only be negative, in a negative potential.

The wavefunction may become infinite at some points, if the potential is infinite there. For instance, for a potential $-\alpha/r^s$ (where $\alpha > 0$ is some constant), the energy of a particle localized over distance a about the origin is $\simeq \hbar^2/ma^2 - \alpha/a^s$; if $s > 2$, the particle falls to the centre; if $s < 2$, the kinetic energy may equilibrate the potential energy, and there may exist bound states (a classical particle falls always to the centre for $s > 0$); the case $s = 2$ requires a special discussion. For $r \to \infty$ we may assume a particle localized in the spherical shell with width $\Delta r \sim r$; for $s < 2$ the energy $\hbar^2/mr^2 - \alpha/r^s$ becomes negative for large r, therefore the particle may be found there, at infinity, with a fair probability; there exist bound states and, more, they are denser towards zero energy. For $s > 2$, the negative spectrum is finite, there exists a finite number of bound states.

We can compute the velocity of the particle, by $\mathbf{v} = \dot{\mathbf{r}} = \frac{i}{\hbar}[H, \mathbf{r}] = \mathbf{p}/m$; since $\mathbf{r}$ and $\mathbf{p}$ have not simultaneously definite values, nor $\mathbf{r}$ and $\mathbf{v}$ have; the acceleration is $\mathbf{a} = \dot{\mathbf{v}} = -(1/m)gradV$, as for a classical particle. From equations (5.1) and (5.3) we get immediately

$$\frac{\partial \mid \psi \mid^2}{\partial t} + \frac{i\hbar}{2m} div \left[\psi^* grad\psi - \psi grad\psi^* \right] = 0 \ , \qquad (5.6)$$

where

$$\mathbf{j} = \frac{i\hbar}{2m} \left[\psi grad\psi^* - \psi^* grad\psi \right] \qquad (5.7)$$

is the particle current density; if $\psi = \mid \psi \mid e^{i\alpha}$, then $\mathbf{j} = \frac{\hbar}{m} \mid \psi \mid^2$ $grad\alpha$; if $\alpha = \mathbf{pr}/\hbar$ (free particle), $\mathbf{j} = (\mathbf{p}/m) \mid \psi \mid^2 = \mathbf{v} \mid \psi \mid^2$; if we normalize the wavefunction such that one particle crosses the unit area perpendicular to the flow, on average, *i.e.* $j = 1$ (unit current density), then the wavefunction is $\psi = (1/\sqrt{v}) e^{-\frac{i}{\hbar}Et + \frac{i}{\hbar}\mathbf{pr}}$. Equation (5.6) is the probability conservation.

Equation (5.2) can be obtained as a stationary-state condition for $\int d\mathbf{r} \psi^* (H - E) \psi$, or $\int d\mathbf{r} \psi^* H \psi$ with the constraint $\int d\mathbf{r} \psi^* \psi = 1$; indeed the variation of this quantity leads to the Schrodinger equation. The minimum value E_0 corresponds to the ground-state wavefunction ψ_0 (normal state); excited states ψ_n (with higher E_n) should be orthogonal to ψ_0; their extremum condition should be seek with the orthogonality constraints. This is called the variational-principle for Schrodinger's equation. It is easy to see that (real) ψ_0 should not have zeros (no nodes), inside the definition domain, but ψ_n should have; the normal state cannot be degenerate, since a combination of wavefunctions may have a zero. Expanding the domain of integration, the energy levels decrease. All these statements can be easily seen by using $(grad\psi)^2$ instead of $\psi \Delta \psi$ (for many-particle wavefunctions such assertions are not valid).

If the potential in the Schrodinger equation is separable, then the one-dimensional motion may be relevant. The free-particle spectrum is doubly degenerate, with $e^{\pm ikx}$, $k = p/\hbar$, corresponding to the motion in either direction. It is easy to see that in one dimension the bound states are not degenerate (the wronskian is vanishing). The variational principle indicates that the n-th bound-state excited level has a wave-function with n nodes inside the region is confined to (this is sometimes called the oscillation theorem). In one dimension the motion may be blocked in one direction and free in the other; in order to normalize the wavefunction we may omit finite-extension regions and use the asymptotic wavefunction, which is of the form $\psi_k \sim \cos(kx+\delta)$ for $x \rightarrow \infty$ and $k = \sqrt{2mE/\hbar^2}$ $(E > 0)$; δ is some constant; it is easy to see that the normalization is $\int dx\psi_k\psi_{k'} = \delta(k-k^{'}) = (1/\hbar v)\delta(E-E^{'})$, with the normalization constant $1/\sqrt{2\pi}$.

5.2 Potential well and potential barrier

Let us assume a one-dimensional potential well $-V_0$ $(V_0 > 0)$ extending from $-a/2$ to $a/2$; Schrodinger's equation for bound states is $\psi^{''} - \kappa^2\psi = 0$ outside the well ($x < -a/2$ and $x > a/2$) and $\psi^{''} + \kappa_1^2\psi = 0$ inside the well ($-a/2 < x < a/2$), where $\kappa^2 = 2m\cdot$ $\cdot \mid E \mid /\hbar^2$, $\kappa_1^2 = 2m(V_0 - \mid E \mid)/\hbar^2$, $\kappa^2 + \kappa_1^2 = \kappa_0^2$, $\kappa_0^2 = 2mV_0/\hbar^2$; $\mid E \mid$ is limited to $0 <\mid E \mid< V_0$ $(\kappa_1^2 > 0)$. The wavefunction is $Ae^{-\kappa x}$ for $x > a/2$, $Be^{\kappa x}$ for $x < -a/2$, $Ce^{i\kappa_1 x} + De^{-i\kappa_1 x}$ for $-a/2 < x < a/2$. This Schrodinger equation is symmetric for the spatial inversion, so we have symmetric solutions $A = B$, $C = D$ and antisymmetric solutions $A = -B$, $C = -D$ (the solutions have a well-determined parity). The continuity of the wavefunction and its derivative at $x = \pm a/2$ leads to $\tan\kappa_1 a/2 = \pm\kappa/\kappa_1$, or, with $\kappa = \kappa_0 \sin\alpha$, $\kappa_1 = \kappa_0\cos\alpha$, $\sin(\kappa_1 a/2 \pm \alpha) = 0$; it follows $\kappa_1 a/2 = \pi n \mp \alpha$, and the energy levels

$$\mid E_n \mid= V_0 - \frac{\pi^2\hbar^2}{ma^2}(n \mp \alpha/\pi)^2 \quad , \tag{5.8}$$

where n is an integer and α, given by $\sin\alpha =\mid E \mid /V_0$ varies between 0 and $\pi/2$; the integer n is limited to some $\pm n_0$, such that $0 <\mid E_n \mid< V_0$. There exists a finite number of energy levels, with the general structure $E_b \simeq V_0 - \hbar^2 n^2/ma^2$, where $E_b =\mid E_n \mid$ is the binding energy of the particle, V_0 is the potential energy and $\hbar^2 n^2/ma^2$ is the kinetic energy; $\hbar^2/ma^2$ is the localization energy of the particle in the well. For a deep well ($V_0 \gg \hbar^2/ma^2$) the energy levels are dense, for a shallow well $V_0 \ll \hbar^2/ma^2$) the particle is not bound. The ground state is located approximately at $\mid E_0 \mid\simeq V_0 - \pi^2\hbar^2/4ma^2$ (> 0); for a shallow well, there may exist only one bound state with energy $\mid E \mid\simeq V_0$. The result can be extended immediately to a three-dimensional rectangular well. It is worth noting that the particle exerts a pressure $\sim \hbar^2 n^2/ma^3$ on the walls.

For $E > 0$ (continuous spectrum) the Schrodinger equation is $\psi'' + k^2\psi = 0$ outside the well and $\psi'' + k_1^2\psi = 0$ inside the well, where $k^2 = 2mE/\hbar^2$, $k_1^2 = 2m(E - V_0)/\hbar^2$; in the region $x \to -\infty$ the wavefunction is asimptotically $e^{ikx} + Ae^{-ikx}$, where $k^2 = 2mE/\hbar^2$; we can see that apart from the incident wave e^{ikx} there exists a reflected wave Ae^{-ikx}. In the region $x \to +\infty$ we may limit ourselves to the transmitted wave Be^{ikx} (as a boundary condition); inside the well the wavefunction is $Ce^{ik_1x} + De^{-ik_1x}$ for $E > V_0$; for $0 < E < V_0$ the wavenumber k_1 is replaced by $i\kappa$, where $\kappa^2 = 2m(V_0 - E)/\hbar^2$. The continuity of the wavefunction and its derivative at $x = \pm a/2$ gives the coefficients A, B, C, D. The incident current density is proportional to k, the reflected current density is proportional to $k \mid A \mid^2$, the transmitted current density is proportional to $k \mid B \mid^2$; the reflection coefficient is $R^2 =\mid A \mid^2$, *i.e.* the ratio of the reflected current to the incident current, while the ratio of the transmitted current to the incident current is the trasmission coefficient $T^2 =\mid B \mid^2$; of course, $R^2 + T^2 = 1$.

We compute the transmission coefficient for a rectangular barrier $V_0 > 0$ extending from $-a/2$ to $a/2$; for $E > V_0$ we get

$$T^2 = \frac{4k^2k_1^2}{(k^2 - k_1^2)^2 \sin^2 ak_1 + 4k^2k_1^2} \ ; \tag{5.9}$$

for $0 < E < V_0$ the wavenumber k_1 in equation (5.9) is replaced by $i\kappa$, where $\kappa^2 = 2m(V_0 - E)/\hbar^2$. We can see that for $E \gg V_0$ the transmission coefficient goes to unity, as expected, while for $E \ll V_0$ the transmission coefficient is exponentially small (and proportional to the energy); for $E \simeq V_0$, the transmission coefficient is $\simeq 1/(1 + a^2k^2/4)$.

5.3 Linear harmonic oscillator

The linear harmonic oscillator is defined by the hamiltonian

$$H = \frac{1}{2m}p^2 + \frac{1}{2}m\omega^2x^2 \ , \tag{5.10}$$

where m is the mass of the particle and ω is the frequency of the oscillator; Schrodinger's equation reads

$$\psi'' + \frac{2m}{\hbar^2}(E - \frac{1}{2}m\omega^2 x^2)\psi = 0 \ ; \tag{5.11}$$

with the substitutions $\xi = \sqrt{m\omega/\hbar}x$, $\psi = e^{-\xi^2/2}\chi$ equation (5.11) becomes

$$\chi'' - 2\xi\chi + (2E/\hbar\omega - 1)\chi = 0 \ . \tag{5.12}$$

We seek the solution as a power series $\chi = \sum_{n=0} a_n \xi^n$; we get

$$(n+2)(n+1)a_{n+2} = -(2E/\hbar\omega - 1 - 2n)a_n \ ; \tag{5.13}$$

finite solutions are obtained by interrupting the series at $n = 0, 1, 2, ...$, *i.e.* for

$$E = E_n = \frac{\hbar\omega}{2}(2n+1) \ ; \tag{5.14}$$

the solutions χ_n are polynomials of powers of ξ; they are the Hermite polynomials

$$H_n(\xi) = (-1)^n e^{\xi^2}\frac{d^n}{d\xi^n}e^{-\xi^2} \ ; \tag{5.15}$$

the wavefunctions are

$$\psi_n(\xi) = \frac{1}{\pi^{1/4}2^{n/2}\sqrt{n!}}e^{-\xi^2/2}H_n(\xi) \tag{5.16}$$

(normalized with respect to ξ); the matrix elements of the coordinate are $x_{n,n-1} = x_{n-1,n} = \sqrt{n\hbar/2\omega}$. We note that equation (5.11) can be written as

$$\left(\frac{d}{d\xi} + \xi\right)\left(\frac{d}{d\xi} - \xi\right)\psi = -(2E/\hbar\omega + 1)\psi \ , \tag{5.17}$$

which shows that ψ_n can be obtained recurrently by applying $d/d\xi - \xi$ and the ground-state wavefunction is given by $(d/d\xi + \xi)\psi_0 = 0$, $\psi_0 \sim e^{-\xi^2/2}$. In addition,

$$\begin{gathered}\left(\tfrac{d}{d\xi} + \xi\right)\psi_n = \sqrt{2n}\psi_{n-1} \ , \\ \left(\tfrac{d}{d\xi} - \xi\right)\psi_n = -\sqrt{2(n+1)}\psi_{n+1} \ .\end{gathered} \tag{5.18}$$

There exists an interesting peculiarity of the harmonic oscillator. Let us form the superposition of wavefunctions

$$\Phi = e^{-\frac{i\omega t}{2}} \sum_{n=0} \left(\frac{A}{2} e^{-i\omega t} \right)^n \frac{1}{n!} e^{-\xi^2/2} H_n(\xi) \quad , \tag{5.19}$$

where the constant A is much greater than unity ($A \gg 1$); we can see that this is a superposition of wavefunctions ψ_n given by equation (5.16) (up to $\pi^{1/4}$), time factor $e^{-iE_n t/\hbar}$ included, with weights $\sqrt{\frac{(A^2/2)^n}{n!}}$; these weights have a pretty sharp peak for large $n \simeq A^2/2$, which means that we include in equation (5.19) mainly those n lying in the vicinity of this large n, *i.e.* we form a wavepacket in equation (5.19). The summation in equation (5.19) can be performed with the generating function of the Hermite polynomials,

$$\sum_{n=0} \frac{s^n}{n!} H_n(\xi) = e^{-s^2+2s\xi} \quad ; \tag{5.20}$$

we get

$$\Phi = e^{A^2/4 - \frac{1}{2}(\xi - A\cos\omega t)^2} e^{-i\left[\frac{1}{2}\omega t + A\sin\omega t\left(\xi - \frac{1}{2}A\cos\omega t\right)\right]} \quad . \tag{5.21}$$

This remarkable result[1] shows that Φ is localized about $\xi = A\cos\omega t$ ($x = \sqrt{\hbar/m\omega}A\cos\omega t$), which is the coordinate of a classical oscillator; the localization is of the order unity (in ξ) and Φ is rather high therein ($\sim e^{A^2/4}$). The energy of the classical oscillator is $E = ma^2\omega^2/2$, where $a = \sqrt{\hbar/m\omega}A$; we get $E = \hbar\omega A^2/2 = n\hbar\omega$ (for $A^2 = 2n$); as if the amplitude $A^2/2$ includes n elementary energy levels $\hbar\omega$ (*e.g.* photons, quantum-field particles). Therefore, a quantum particle whose motion consists of (includes) approximately n quanta (of harmonic oscillator), with large n, behaves like a classical particle. The analogy implies also the oscillations included in equation (5.21), which are numerous (A is large), more numerous (and sharper) near the centre $\xi = 0$ and vanishing near $\xi = \pm A$. Moreover, and this is indeed remarkable, the wavepacket given by equation (5.21) does not spread

[1] E. Schrodinger, "Der stetige Ubergang von der Mikro- zur Makromechanik", Naturwissenschaften **28** 664 (1926) ("The continuous transition from micro- to macro-mechanics", in *Collected Papers on Wave Mechanics*, Blackie&Son Ltd., London and Glasgow (1928)).

out; it is a sharp, oscillating, compact wavepacket. This is so because we include in Φ discrete levels, which are equidistant. A constant added to the energy levels (like a rest energy of the particle), only makes the oscillations inside the wavepacket faster, without affecting the oscillations of the group. The wavefunction Φ may serve as a construction of classical particles from quanta; it is easy to see that it holds in two or three dimensions too; it holds only for harmonic oscillators (or states of identical particles), where, the superposition in equation (5.19) must be made with the same phase. This common-phase feature defines the so-called coherent states; a coherent state is a quantum state which behaves like a special classical particle (or field).

5.4 Angular momentum

A function ψ of coordinates $\mathbf{r}_a$ of several particles (denoted by a) changes under an infinitesimal translation $b_a \rightarrow \mathbf{r}_a + \delta\mathbf{r}$ as

$$\psi(...\mathbf{r}_a + \delta\mathbf{r}...) = \left[1 + \delta\mathbf{r}\sum_a grad_a + ...\right]\psi(...\mathbf{r}_a...) \quad ; \qquad (5.22)$$

for a quasi-classical function $\psi = e^{\frac{i}{\hbar}S}$, where S is the mechanical action, the operator $1 + \delta\mathbf{r}\sum_a grad_a$ of the infinitesimal translations is

$$T = 1 + \frac{i}{\hbar}\delta\mathbf{r}\sum_a \mathbf{p}_a \quad , \qquad (5.23)$$

where $\mathbf{p}_a = -i\hbar grad_a$ is the momentum operator and $\mathbf{p} = \sum_a \mathbf{p}_a$ is the total momentum; indeed, it changes the wavefunction by the phase $e^{\frac{i}{\hbar}\mathbf{p}\delta\mathbf{r}}$, which is a constant if the space is homogeneous, *i.e.* $\mathbf{p}$ is conserved; in which case, it must commute with the hamiltonian; the assembly of particles must be closed, *i.e.* external forces must be absent. Similarly, for an infinitesimal rotation $\delta\mathbf{r} = \delta\boldsymbol{\alpha} \times \mathbf{r}_a$ with angle $\delta\boldsymbol{\alpha}$ we get the operator

$$R = 1 + \frac{i}{\hbar}\delta\boldsymbol{\alpha}\sum_a \mathbf{r}_a \times \mathbf{p}_a \quad , \qquad (5.24)$$

where $\mathbf{r}_a \times \mathbf{p}_a$ is the angular momentum of the a-th particle; we denote it by $\hbar\mathbf{L} = \sum_a \mathbf{r}_a \times \mathbf{p}_a$, and call $\mathbf{L}$ angular momentum; it is measured in units $\hbar$. For an isotropic space, like a closed assembly, $\mathbf{L}$ is conserved; however, the space is isotropic for a centrally symmetric field too, *i.e.* for a potential which depends only on r_a; the angular momentum is conserved in that case (for each particle); $\mathbf{L}$ changes the wavefunction by the phase factor $e^{i\mathbf{L}\delta\boldsymbol{\alpha}}$. In non-degenerate stationary states the mean value of the angular momentum is zero, since it changes the sign under time reversal (while the state remains unchanged).

The commutation relations of the angular momentum are $[L_i, L_j] = i\varepsilon_{ijk}L_k$, where i, j, k are cartesian components and ε_{ijk} is the antisymmetric unit tensor of rank three; we get $[L_i, L^2] = 0$, where $L^2 = L_i^2$ (with summation over i); similarly, $[L_i, x_j] = i\varepsilon_{ijk}x_k$ and $[L_i, p_j] = i\varepsilon_{ijk}p_k$; only L^2 and a component of $\mathbf{L}$ can be measured simultaneously. It is convenient to use $L_\pm = L_x \pm L_y$ and $[L_+, L_-] = 2L_z$, $[L_z, L_\pm] = \pm L_\pm$, $L^2 = L_+L_- + L_z^2 - L_z$. If $L_z\varphi_m = m\varphi_m$ is the equation for eigenvalues and eigenfunctions of the L_z, then $L_z(L_\pm\varphi_m) = (m \pm 1)(L_\pm\varphi_m)$ and there must by some limits $\pm l$ for the eigenvalues m, since $L^2 - L_z^2 > 0$; since $L_+\varphi_l = 0$ (and $L\varphi_{-l} = 0$), it follows $L^2\varphi_l = l(l+1)$, a relation which is preserved for all m from $-l$ to l; therefore, we have eigenfunctions φ_{lm} and $2l + 1$ values of m for any fixed l; since m changes by unity, it follows that $2l+1$ must be an integer; *i.e.*, l must be a positive integer or a positive half-integer. From the above commutation relations we get the matrix elements

$$(L_x)_{m,m-1} = (iL_y)_{m,m-1} = \frac{1}{2}\sqrt{(l+m)(l-m+1)} \ . \tag{5.25}$$

In spherical coordinates $L_z = -i\partial/\partial\varphi$; its eigenfunctions are $e^{im\varphi}$; since they must be single-valued, m and l must be integers. The other components are

$$\begin{aligned} L_\pm &= e^{\pm i\varphi}\left(\pm\frac{\partial}{\partial\theta} + i\cot\theta\frac{\partial}{\partial\varphi}\right) \ , \\ L^2 &= -\frac{1}{\sin\theta}\frac{\partial}{\partial\theta}\left(\sin\theta\frac{\partial}{\partial\theta}\right) - \frac{1}{\sin^2\theta}\frac{\partial^2}{\partial\varphi^2} \ . \end{aligned} \tag{5.26}$$

The (normalized) eigenfunctions are the spherical surface harmonics defined by

$$Y_{lm}(\theta,\varphi) = (-1)^{(m+|m|)/2} i^l \cdot \left[\frac{2l+1}{4\pi}\frac{(l-|m|)!}{(l+|m|)!}\right]^{1/2} P_l^{|m|}(\cos\theta)e^{im\varphi} \quad , \tag{5.27}$$

where P_l^m $(0 \leq m \leq l)$ are the associated Legendre polynomials:

$$P_l^m = (-1)^l \frac{\sin^m\theta}{2^l l!}\frac{d^{l+m}}{d(\cos\theta)^{l+m}}\sin^{2l}\theta \ ; \tag{5.28}$$

they satisfy the equation

$$\frac{1}{\sin\theta}\frac{d}{d\theta}\left(\sin\theta\frac{dP_l^m}{d\theta}\right) - \frac{m^2}{\sin^2\theta}P_l^m + l(l+1)P_l^m = 0 \ . \tag{5.29}$$

There exists the relationship $(-1)^{l-m}Y_{l,-m} = Y_{lm}^*$; Y_{00} is a constant (spherical symmetry); for fixed m there exist $l- \mid m \mid$ eigenvalues of L^2; consequently, Y_{lm} have $l- \mid m \mid$ zeros (nodal "lines of latitude") for $l- \mid m \mid$ different values of θ; if $\cos m\varphi$ or $\sin m\varphi$ are used, there exist m nodal "lines of longitude".

Two independent angular momentum $\mathbf{L}_{1,2}$ can be added to $\mathbf{L} = \mathbf{L}_1 + \mathbf{L}_2$; the eigenfunctions, defined by l and m are linear combinations of products of eigenvectors defined by l_1, m_1 and l_2, m_2, a total number of $(2l_1+1)(2l_2+1)$ eigenvectors. Since $m = m_1 + m_2$, the maximum m is $L = l_1 + l_2$; the next m is $L - 1 = l_1 + l_2 - 1$, which arises from $l_1 - 1$ or $l_2 - 1$; it belongs either to the set $l_1 + l_2$, or to the set $l_1 + l_2 - 1$; we continue, and find out that the maximum L is $l_1 + l_2$ and the minimum L is $\mid l_1 - l_2 \mid$. This corresponds to the classical addition of two vectors. The coefficients of the linear combinations of the products of angular-momentum eigenfunctions are called Clebsch-Gordan coefficients (or Wigner $3j$-symbols).[2] Summations of three angular momenta imply coefficients called $6j$-symbols, which may be generalized to even higher-order $9j$-symbols, etc.

[2] M. Born, W. Heisenberg and P. Jordan, "Zur Quantenmechanik II", Z. Phys. **35** 557 (1926); M. Born and P. Jordan, *Elementare Quantenmechanik*, Springer (1930); A. R. Edmonds, *Angular Momentum in Quantum Mechanics*, Princeton University Press, Princeton (1957).

5.5 Symmetries

A change in the wavefunction ψ can be represented by an operator O: $\psi \to O\psi$; for instance, O can be the operator of the infinitesimal translations or the operator of the infinitesimal rotations derived above. If an operator F has the same eigenvalue f for ψ and $O\psi$, then O represents a symmetry for that operator; this means $FO\psi = fO\psi = OF\psi$, *i.e.* O and F must commute: $[O, F] = 0$. In particular, for translations O reduces to the total momentum $\mathbf{p}$, for rotations O reduces to the total angular momentum $\mathbf{L}$ (here, orbital momentum); $\mathbf{p}$ and $\mathbf{L}$ are called the generators of the translation and rotation symmetry, respectively. Any O (of continuous symmetries) may be represented by e^{iG}, where G is the generator. There exists an internal angular momentum $\mathbf{s}$, which is the spin, with half-integers values; the total angular momentum is $\mathbf{J} = \mathbf{L} + \mathbf{s}$. If the hamiltonian H is invariant under a symmetry operation defined by O (*i.e.* G), *i.e.* if $[O, H] = 0$ ($[G, H] = 0$), then the motion is symmetric under that operation; it conserves the eigenvalues of the symmetry generators. The conservation of the symmetries eigenvalues may cancel out some matrix elements and forbid the corresponding transitions (this circumstance is called selection rule).

A particular symmetry is the spatial inversion (which is discrete); some vectors, like the position, the momentum, change the sign under this operation; they are called polar vectors; some others, like the angular momentum, or the magnetic field, do not; they are called axial vectors. Some scalars are invariant under spatial inversion, they are called true scalars; others, like the scalar product of a polar vector and an axial vector, change the sign under spatial inversion; they are called pseudoscalars. The eigenvalues of the generator P of the spatial inversion are ± 1, since a double application of P leaves the wavefunction unchanged. P is also called the parity operator, and its eigenfunctions are even or odd; they have the parity $+1$ or -1, respectively. In spherical coordinates the angles change according to $\varphi \to \pi = \varphi$, $\theta \to \pi - \theta$ under spatial inversion; the spherical harmonics Y_{lm} given by equation (5.27) have the parity $(-1)^l$.

An operator F changes like OFO^{-1} under a symmetry operation O, since $OF\psi = OFO^{-1}O\psi$ (actually OFO^{+}, since O is unitary); if it is conserved, $[O, F] = 0$ ($[G, F] = 0$). Under rotations, the tensors become linear combinations with the rotation-matrix elements.

A scalar f (under rotations) is invariant; therefore, $[\mathbf{L}, f] = 0$; f is diagonal in m and does not depend on m; its matrix elements are $f_{n'lm,nlm} = f_{n'l,nl}$, where n, n' are other quantum numbers (corresponding to other operators in the complete set of measurable quantities).

The components of a vector $\mathbf{A}$ become linear combinations under rotations; therefore, $[L_i, A_j]$ is a linear combination of A_k; the coefficients are ε_{ijk}, $[L_i, A_j] = \varepsilon_{ijk}A_k$, according to the commutation rules of L_i and x_k, or p_k. This commutation relation may help to get the matrix elements of $\mathbf{A}$; it is worth noting that they imply also transitions $l \rightarrow l \pm 1$. They are given by

$$\begin{aligned}
(A_z)_{n'lm,nlm} &= \frac{m}{\sqrt{l(l+1)(1l+1)}}A_{n'l,nl} \ , \\
(A_z)_{n'lm,nl-1m} &= \sqrt{\frac{l^2-m^2}{l(2l-1)(2l+1)}}A_{n'l,nl-1} \ , \\
(A_z)_{n'l-1m,nlm} &= \sqrt{\frac{l^2-m^2}{l(2l-1)(2l+1)}}A_{n'l-1,nl}
\end{aligned} \tag{5.30}$$

and

$$\begin{aligned}
(A_-)_{n'lm-1,nlm} &= \sqrt{\frac{(l-m+1)(l+m)}{l(l+1)(2l+1)}}A_{n'l,nl} \ , \\
(A_-)_{n'lm-1,nl-1m} &= \sqrt{\frac{(l-m+1)(l-m)}{l(2l-1)(2l+1)}}A_{n'l,nl-1} \ , \\
(A_-)_{n'l-1m-1,nlm} &= \sqrt{\frac{(l+m-1)(l+m)}{l(2l-1)(2l+1)}}A_{n'l-1,nl} \ ;
\end{aligned} \tag{5.31}$$

the denominators are conventional; $A_{n'l',nl}$ are called reduced matrix elements; equations (5.30) and (5.31) are called the Eckart-Wigner theorem; $(A_+)_{n'l'm',nlm} = (A_-)^*_{nlm,n'l'm'}$. The scalar product has the matrix elements

$$(\mathbf{AB})_{n'lm,nlm} = \frac{1}{2l+1}\sum_{n''l''} A_{n'l,n''l''}B_{n''l'',nl} \ . \tag{5.32}$$

Comparing equations (5.30) and (5.31) with equations (5.25), we get the reduced matrix elements

$$L_{l,l} = \sqrt{l(l+1)2l+1} \ , \ L_{l-1,l} = L_{l,l-1} = 0 \ . \tag{5.33}$$

By using the spherical harmonics we may get $(\cos\theta)_{l-10,l0} = il/$ $/\sqrt{(2l-1)(2l+1)}$; and its reduced matrix elements $n_{l-1,l} = -n_{l,l-1} = i\sqrt{l}$, $n_{l,l} = 0$, where $\mathbf{n}$ is the unit vector along the position $\mathbf{r}$.

The mean value of $n_i n_k - \frac{1}{3}\delta_{ik}$ is of the form $a(L_iL_k + L_kL_i - \frac{2}{3}\delta_{ik})$; by using the commutation relations, $a = -1/(2l-1)(2l+3)$.

The spherical harmonics given by equation (5.27) for $l = 1$ can be written as

$$\begin{aligned} Y_{1\pm1} &= \mp i\sqrt{\tfrac{3}{8\pi}} \sin\theta e^{\pm i\varphi} = \mp i\sqrt{\tfrac{3}{8\pi}}(n_x \pm n_y) \ , \\ Y_{10} &= i\sqrt{\tfrac{3}{4\pi}} \cos\theta = i\sqrt{\tfrac{3}{4\pi}} n_z \ , \end{aligned} \tag{5.34}$$

where $\mathbf{n}$ is the unit vector along the position $\mathbf{r}$; it follows that, as regards the transformation properties under rotations, any vector $\mathbf{A}$ can be related to the eigenfunctions ψ_{1m} of the angular momentum $L = 1$:

$$\psi_{1\pm1} = \mp\frac{i}{\sqrt{2}}(A_x \pm A_y) \ , \ \psi_{10} = iA_z \ ; \tag{5.35}$$

therefore, any vector behaves as the angular momentum $\mathbf{L}$, where $L = 1$.

If we carry out a rotation on the eigenfunctions of the angular momentum, the rotated eigenfunctions are obtained from the old ones as linear combinations; indeed, the rotated functions are obtained by acting with the operators of the angular momentum; for spherical harmonics we can check this property directly; since the angular-momentum eigenfunctions ψ_{jm} are transformed between themselves under rotations, and cannot be decomposed into smaller sub-sets with indepenent transformations, the set of eigenfunctions ψ_{jm} is said to form an irreducible representation of the rotation group (the rotations form a group, called $SO(3)$). On the other hand, we can see that the $2j+1$ eigenfunctions ψ_{jm} form a tensor of rank j; it is called a spherical tensor. A vector is a spherical tensor of rank one. If we add two angular momenta j_1 and j_2 to the angular momentum j, the eigenfunctions ψ_{jm} are linear combinations of products $\psi_{j_1m_1}\psi_{j_2m_2}$; it follows that products like $\psi_{j_1m_1}\psi_{j_2m_2}$ can be written as linear combinations of ψ_{jm}; now, if we multiply by $\psi^*_{j_3m_3}$ and integrate over angles, we get the matrix elements of the tensor $\psi_{j_im_1}$, which corresponds to $j_3 = j$; therefore, we can see that the matrix elements of a spherical tensor should obey the rules of addition of angular momenta, *e.g.* j_3 runs from $j_1 + j_2$ to $\mid j_1 - j_2 \mid$. This is the origin of the selection rule $l \to l \pm 1$ (and $l \to l$) for $j_1 = 1$ (a vector).

5.6 Centrally symmetric field

The hamiltonian

$$H = -\frac{\hbar^2}{2m_1}\Delta_1 - \frac{\hbar^2}{2m_2}\Delta_2 + V(\mid \mathbf{r}_1 - \mathbf{r}_2 \mid) \tag{5.36}$$

of two particles with mass $m_{1,2}$ interacting by $V(\mid \mathbf{r}_1 - \mathbf{r}_2 \mid)$, where $\mathbf{r}_{1,2}$ are the positions of the particles, can be separated in the hamiltonian of the centre of mass and the hamiltonian of the relative motion,

$$H = -\frac{\hbar^2}{2M}\Delta_R - \frac{\hbar^2}{2m}\Delta_r + V(r) \ , \tag{5.37}$$

where $\mathbf{R} = (m_1\mathbf{r}_1 + m_2\mathbf{r}_2)/M$, $\mathbf{r} = \mathbf{r}_1 - \mathbf{r}_2$, $M = m_1 + m_2$ and $m = m_1m_2/M$; the center-of-mass motion is free, the relative motion is a motion in a centrally symmetric potential. In spherical coordinates, the hamiltonian of the relative motion becomes

$$H = -\frac{\hbar^2}{2m} \cdot \frac{1}{r^2}\frac{\partial}{\partial r}\left(r^2\frac{\partial}{\partial r}\right) + \frac{\hbar^2 L^2}{2mr^2} + V(r) \ , \tag{5.38}$$

where L^2 is given by equation (5.26). If we introduce the radial momentum $p_r\psi = -i\hbar(1/r)\partial(r\psi)$, the hamiltonian becomes

$$H = \frac{1}{2m}p_r^2 + \frac{\hbar^2 L^2}{2m} + V(r) \ . \tag{5.39}$$

The hamiltonian given by equation (5.38) is separable, so we seek the wavefunctions as $\psi = R(r)Y_{lm}$, where Y_{lm} are the spherical harmonics;

the radial Schrodinger equation becomes

$$\frac{1}{r^2}\frac{d}{dr}\left(r^2\frac{dR}{dr}\right)-\frac{l(l+1)}{r^2}R+\frac{2m}{\hbar^2}\left[E-V(r)\right]R=0\ ; \qquad (5.40)$$

the energy levels are degenerate with respect to m ($2l+1$-fold degeneracy); with the substitution $R=\chi/r$, equation (5.40) becomes

$$\chi''-\frac{l(l+1)}{r^2}\chi+\frac{2m}{\hbar^2}\left[E-V(r)\right]\chi=0\ ; \qquad (5.41)$$

we seek solutions with the boundary condition $\chi(0)=0$, which precludes the fall to the centre. In one dimension, the states vanishing in one direction are not degenerate (their wronskian is zero); therefore, the states here are determined entirely by E, l and m; the corresponding operators form a complete set of observables. We note that the motion in equation (5.41) proceeds in an effective potential $V_l=V+\hbar^2 l(l+1)/2mr^2$; the latter part of this potential is the centrifugal potential; it increases the energy. We can classify the states by l, m and the radial number n_r of nodes of the radial wavefunctions; the states with $l=0,\,1,\,2,\,3,...$ are also denoted by s, p, d, f,...; the ground state is an s-state (no nodes).

If $r^2V(r)\rightarrow 0$ for $r\rightarrow 0$, then $\chi\sim r^s$ for $r\rightarrow 0$, where $s=l+1$; therefore, $R\sim r^l$ for $r\rightarrow 0$, and the probability of localization of the particle near the origin is $\sim r^{2(l+1)}$; it decreases with increasing l.

5.7 Spherical waves

Let us consider the Schrodinger equation in spherical coordinates for the free motion; the wavefunctions are RY_{lm}, where R satisfies the equation

$$R''+\frac{2}{r}R'+\left[k^2-\frac{l(l+1)}{r^2}\right]R=0\ ,\ \ k^2=2mE/\hbar^2\ ; \qquad (5.42)$$

the spectrum is continuous ($E>0$) and R is labelled by k and l. By the substitution $R_{kl}=r^l\chi_{kl}$, we get

$$\chi_{kl}''+\frac{2(l+1)}{r}\chi_{kl}'+k^2\chi_{kl}=0\ ; \qquad (5.43)$$

taking the derivative of this equation we notice that $\chi'_{kl}/r \sim \chi_{kl+1}$; therefore $R_{kl} \sim r^l \left(\frac{1}{r}\frac{d}{dr}\right)^l R_{k0}$. For $l = 0$ we get $R'' + 2R'/r + k^2 R = 0$, whence $R_{k0} \sim \frac{1}{r}\sin kr$; we choose

$$R_{kl} = (-1)^l \frac{2r^l}{k^l}\left(\frac{1}{r}\frac{d}{dr}\right)^l \frac{\sin kr}{r} \ ; \tag{5.44}$$

these functions are normalized by

$$\int_0^\infty dr \cdot r^2 R_{kl} R_{k'l'} = 2\pi\delta(k - k') \ ; \tag{5.45}$$

they are, in fact, the spherical Bessel functions $R_{kl} = 2kj_l(kr)$, $j_l(z) = \sqrt{\pi/2z}J_{l+1/2}(z)$, where $J_{l+1/2}$ are the Bessel functions of half-integral order. The asymptotic behaviour of these functions is obtained easily from equation (5.44); is

$$R_{kl} \sim \frac{2\sin(kr - l\pi/2)}{r} \quad kr \gg 1 \ ; \tag{5.46}$$

the normalization can also be obtained immediately by using this asymptotic behaviour. Near the origin, the functions R_{kl} behave like

$$R_{kl} \sim \frac{2k^{l+1}}{(2l+1)!!} r^l \quad kr \ll 1 \ . \tag{5.47}$$

We can also use travelling spherical waves, ingoing $(-)$ or outgoing $(+)$ (they are suitable for scattering problems); these functions are given by

$$R^{\pm}_{kl} = (-1)^l A \frac{r^l}{k^l}\left(\frac{1}{r}\frac{d}{dr}\right)^l \frac{e^{\pm i' kr}}{r} \ ; \tag{5.48}$$

they are related to the Hankel functions $H^{\pm}_{l+1/2}$ by

$$R^{\pm}_{kl} = \pm iA\sqrt{k\pi/2r}H^{(1,2)}_{l+1/2}(kr) \tag{5.49}$$

(with + for 1 and − for 2). The asymptotic expressions are

$$R^{\pm}_{kl} \sim A\frac{e^{\pm i(kr - l\pi/2)}}{r} \quad kr \gg 1 \tag{5.50}$$

and

$$R^{\pm}_{kl} \sim A\frac{(2l-1)!!}{k^l} r^{-l-1} \quad kr \ll 1 \ . \tag{5.51}$$

The normalization of these wavefunctions proceeds as follows: first we note that at large distance the spherical waves may be approximated by plane waves, and the current is $v\psi^*\psi$, where $v = \hbar k/m$ is the velocity; making use of the asymptotic behaviour, we get $A = 1/\sqrt{v}$.

If the field falls off faster than $1/r$, then we may neglect it in equation (5.42) (and the centrifugal field); equation (5.42) becomes

$$\frac{1}{r}\frac{d^2}{dr^2}(rR_{kl}) + k^2 R_{kl} = 0 \ , \tag{5.52}$$

with the general solution

$$R_{kl} \simeq \frac{2\sin(kr - l\pi/2 + \delta_{kl})}{r} \ , \tag{5.53}$$

where δ_{kl} is a phase shift; when the field is absent $\delta k_l = 0$; it can be determined by solving the exact Schrodinger equation. These phase shifts are related to the scattering process. They give also a characterization of the continuous spectrum in potentials with central symmetry.

Spherical potentials wells or spherical oscillator can be considered in spherical coordinates. In general, there exist inversions of l with respect to the increasing radial number in the order of the energy levels.

In cartesian coordinates the wavefunction of the free particle (equation (5.42)) is the plane wave; let it be e^{ikz}; it is a combination of spherical waves, with $m = 0$; using equation (5.44) and comparing the terms, we get

$$e^{ikz} = \sum_{l=0}(-i)^l(2l+1)P_l(\cos\theta)\left(\frac{r}{k}\right)^l\left(\frac{1}{r}\frac{d}{dr}\right)^l\frac{\sin kr}{kr} \ , \tag{5.54}$$

or

$$\begin{aligned} e^{ikz} &= \tfrac{1}{2k}\textstyle\sum_{l=0} i^l(2l+1)R_{kl}(r)P_l(\cos\theta) = \\ &= \textstyle\sum_{l=0} i^l(2l+1)j_l(kr)P_l(\cos\theta) \ , \end{aligned} \tag{5.55}$$

where $P_l(\cos\theta)$ is the Legendre polynomial of the l-th order; the asymptotic behaviour is

$$e^{ikz} \sim \frac{1}{kr}\sum_{l=0} i^l(2l+1)P_l(\cos\theta)\sin(kr - l\pi/2) \ . \tag{5.56}$$

making use of the addition theorem for spherical harmonics, we get

$$e^{i\mathbf{kr}} = 4\pi \sum_{l=0}\sum_{m=-l}^{l} i^l j_l(kr) Y^*_{lm}(\mathbf{k}/k) Y_{lm}(\mathbf{r}/r) \ . \tag{5.57}$$

If we define the in- and outgoing wavefunctions by $\psi^{\pm}_{klm} = R^{\pm}_{kl}(r)\cdot$ $\cdot Y_{lm}(\theta,\varphi)$, then

$$\psi = e^{ikz} = \sum_{l=0}\sqrt{\pi(2l+1)}\frac{1}{ik}(\psi^+_{kl0} - \psi^-_{kl0}) \ ; \tag{5.58}$$

we can define a cross-section $\sigma_l = \pi(2l+1)/k^2$, and a total cross-section $\sigma \simeq 2\pi l\Delta l/k^2 = 2\pi\hbar^2 l\Delta l/p^2$; making use of the classical angular momentum $\hbar l = \rho p$, where ρ is the impact parameter, we get $\sigma = 2\pi\rho\Delta\rho$, which is indeed the classical cross-section; for large l the particle behaves classically.

5.8 Fall to the centre

If the potential is $V(r) = -\beta/r^s$ near the origin ($\beta > 0$), the particle has stationary states for $s < 2$ and falls to the centre (to the origin) for $s > 2$. For $s = 2$, the Schrodinger equation (for $r \sim 0$) is

$$R'' + 2R'/r + \gamma R/r^2 = 0 \ , \tag{5.59}$$

where $\gamma = 2m\beta/\hbar^2 - l(l+1)$; the energy is supposed finite; also, we assume $\gamma > 0$. Let us assume $R \sim r^s$, where $s(s+1) + \gamma = 0$; *i.e.* $s_{1,2} = -1/2 \pm \sqrt{1/4 - \gamma}$. For $0 < \gamma < 1/4$, these roots are negative and $s_1 > s_2$. The solution is of the form $Ar^{s_1} + Br^{s_2}$. We may assume the potential $-\beta/r_0^2$ for $r < r_0$, where r_0 is a cutoff; in this region the solution is $\sim \frac{1}{r}\sin kr$, where $k = \sqrt{\gamma/r_0}$; the continuity conditions give $B = 0$ for $r_0 \to 0$, which means that we have the solution r^{s_1}, $0 - 1/2 < s_1 < 0$; this solution gives an integrable probability density at the origin; the motion has stationary states in this case. For $1/4 < \gamma$, the continuity conditions give $B/A \sim r_0 e^{i\sqrt{4\gamma-1}}$ and the solution is $R \sim r^{-1/2}\cos[\sqrt{\gamma - 1/4}\ln(r/r_0) + const]$; for $r_0 \to 0$ it oscillates indefinitely and the ground state has the energy $-\infty$; the particle falls to the centre.

The critical value of the field is given by $\gamma = 1/4$; for $l = 0$, we get $V_{cr} = -\hbar^2/8mr^2$.

Let us suppose now that the field goes like $-\beta/r^2$ at large distances. For $0 < \gamma < 1/4$ the solution is Ar^{s_1} for zero energy; it has no nodes; therefore, there exists a finite number of nodes at finite distance, and a finite number of energy levels. On the contrary, for $\gamma > 1/4$, the wavefunction has the oscillatory form given above, which has an infinite number of zeros; in this case there exists an infinite number of energy levels.

If the field is $-\beta/r^2$ for all r, the particle falls for $\gamma > 1/4$ and has no energy levels for $\gamma < 1/4$, since the wavefunction is of the form r^{s_1} everywhere, and has no zeros.

5.9 Coulomb field

The radial Schrodinger equation in the Coulomb field $V(r) = -\alpha/r$ reads

$$R'' + 2R'/r - \frac{l(l+1)}{r^2}R + \frac{2m}{\hbar^2}\left(E + \frac{\alpha}{r}\right)R = 0 \; ; \tag{5.60}$$

first, we consider the attractive case $\alpha > 0$; for a hydrogen-like atom with charge Z the parameter $\alpha = Ze^2$, where $e = -4.8 \times 10^{-10} esu$ is the electron charge. It is convenient to introduce the Bohr radius $a_H = \hbar^2/m\alpha = 0.53/Z(\text{Å})$ (electron mass $m \simeq 10^{-27}g$, Planck's constant $\hbar \simeq 10^{-27} erg \cdot s$) and measure the energy in units $m\alpha^2/\hbar^2 = \alpha/a_H = 27.2Z^2(eV)$ ($1eV = 1.6 \times 10^{-12} erg$; $13.6eV$ is also called *rydberg*); these are called atomic units; equation (5.60) becomes

$$R'' + \frac{2}{r}R' - \frac{l(l+1)}{r^2}R + 2\left(E + \frac{1}{r}\right)R = 0 \; . \tag{5.61}$$

We discuss first the discrete spectrum, and introduce the notations $E = -1/2n^2$, $\rho = 2r/n$; equation (5.61) becomes

$$R'' + \frac{2}{\rho}R' + \left[-\frac{1}{4} + \frac{n}{\rho} - \frac{l(l+1)}{\rho^2}\right] R = 0 \ . \tag{5.62}$$

The asymptotic behaviour suggests the substitution $R = \rho^l e^{-\rho/2} w$; equation (5.62) becomes

$$\rho w'' + (2l + 2 - \rho)w' + (n - l - 1)w = 0 \ ; \tag{5.63}$$

the solution is the (confluent) hypergeometric function $w = F(-n + l + 1, 2l + 2, \rho)$; it is a series of the form $w = \sum_{N=0} a_n \rho^N$, which reduces to a polynom (being therefore finite) for integers $n \geq l + 1$; the energy levels are

$$E_n = -1/2n^2 \rightarrow -\frac{m\alpha^2}{2\hbar^2 n^2} = -\frac{\alpha}{2a_H n^2} \ ; \tag{5.64}$$

this is the well-known Bohr spectrum; n is called the principal quantum number, $n_r = n - l - 1$ is the radial quantum number, there exists $\sum_{l=0}^{n-1}(2l+1) = n^2$ energy levels, which are degenerate with respect to l (Coulomb degeneracy) and m (in the spherical wavefunctions Y_{lm}). The radial wavefunctions can also be written as

$$R_{nl} = -\frac{2}{n^2}\sqrt{\frac{(n-l-1)!}{[(n+l)!]^3}} e^{-r/n} \left(\tfrac{2r}{n}\right)^l L_{n+l}^{2l+1}(2r/n) =$$

$$= \frac{2}{n^{l+2}(2l+1)!}\sqrt{\frac{(n+l)!}{(n-l-1)!}}(2r)^l e^{-r/n}. \tag{5.65}$$

$$\cdot F(-n + l + 1, 2l + 2, 2r/n) \ ,$$

where L_{n+l}^{2l+1} are the generalized Laguerre polynomials; they are normalized by

$$\int_0^\infty dr \cdot r^2 R_{nl}^2 = 1 \ . \tag{5.66}$$

The Laguerre polynomials are given by

$$L_n(z) = e^z \tfrac{d^n}{dz^n}(z^n e^{-z}) \ ,$$

$$L_n^m(z) = \tfrac{n!}{(n-m)!} e^z \tfrac{d^n}{dz^n}(z^{n-m} e^{-z}) = \tag{5.67}$$

$$= (-1)^m \tfrac{n!}{(n-m)!} z^{-m} e^z \tfrac{d^{n-m}}{dz^{n-m}}(z^n e^{-z}) \ .$$

The asymptotic behaviour is

$$R_{nl} \sim r^l \frac{2^{l+1}}{n^{2+l}(2l+1)!}\sqrt{\frac{(n+l)!}{(n-l-1)!}} \ , \ r \to 0 \tag{5.68}$$

and

$$R_{nl} \sim (-1)^{n-l-1} \tfrac{2^n}{n^{l+1}\sqrt{(n+l)!(n-l-1)!}} \cdot$$

$$\cdot r^{n-1} e^{-r/n} \ , \ r \to \infty \ ; \tag{5.69}$$

we can see that the ground-state wavefunction decreases exponentially beyond $r \simeq 1$, *i.e.* $r \simeq a_H$. The mean values of powers of r are

$$\overline{r} = \tfrac{1}{2}[3n^2 - l(l+1)] \ , \ \overline{r^2} = \tfrac{1}{2}n^2[5n^2 + 1 - 3l(l+1)] \ ,$$

$$\overline{r^{-1}} = 1/n^2 \ , \ \overline{r^{-2}} = 1/n^3(l+1/2) \ . \tag{5.70}$$

For positive energies the spectrum is continuous; the wavefunctions can be obtained from those of the discrete spectrum by the formal substitution $n = -i/k$, $\rho = 2ikr$, $E = k^2/2$; the radial wavefunctions are

$$R_{kl} = \frac{C_{kl}}{(2l+1)!}(2kr)^l e^{-ikr} F(i/k + l + 1, 2l+2, 2ikr) \ , \tag{5.71}$$

where $C_{kl} = 2ke^{\pi/2k} \mid \Gamma(l+1-i/k) \mid$, Γ being Euler's gamma function; the normalization is

$$\int_0^\infty dr \cdot r^2 R_{kl} R_{k'l} = 2\pi\delta(k - k') \ . \tag{5.72}$$

The asymptotic behaviour is '

$$R_{kl} \sim \frac{2}{r}\sin\left(kr + \frac{1}{k}\ln(2kr) - l\pi/2 + \delta_l\right) \ , \ r \to \infty \ , \tag{5.73}$$

in accordance with equation (5.53), where $\delta_l = arg\Gamma(l+1-i/k)$ and the ln is immaterial. For $E = 0$

$$F(i/k, 2l+2, 2ikr) = (2l+1)!(2r)^{-l-1/2}J_{2l+1}(\sqrt{8r}) \tag{5.74}$$

and $C_{kl}\sqrt{8\pi}k^{-l+1/2}$; the asymptotic behaviour is

$$R_{kl}/\sqrt{k} \sim (8/r^3)^{1/4}\sin(\sqrt{8r} - l\pi - \pi/4) \ . \tag{5.75}$$

For a repulsive Coulomb potential

$$R_{kl} = \frac{C_{kl}}{(2l+1)!}(2kr)^l e^{ikr} F(i/k + l + 1, 2l+2, -2ikr) \ , \tag{5.76}$$

$C_{kl} = 2ke^{-\pi/2k} \mid \Gamma(l+1+i/k) \mid$ and

$$R_{kl} \sim \frac{2}{r}\sin\left(kr - \frac{1}{k}\ln(2kr) - l\pi/2 + \delta_l\right) \ , \ r \to \infty \ , \tag{5.77}$$

where $\delta_l = arg\Gamma(l+1+i/k)$.

5.10 Coulomb degeneracy

The Lenz vector $\mathbf{A} = \mathbf{r}/r - \mathbf{p} \times \mathbf{l}$ is conserved in classical motion in attractive Coulomb field; its quantum-mechanical counterpart is $\mathbf{A} = \mathbf{r}/r - \frac{1}{2}(\mathbf{p} \times \mathbf{l} - \mathbf{l} \times \mathbf{p})$ (it commutes with the hamiltonian $H = p^2/2 - 1/r$). The commutation relations

$$[l_i, A_j] = i\varepsilon_{ijk}A_k \ , \ [A_i, A_j] = -2iH\varepsilon_{ijk}l_k \tag{5.78}$$

can be obtained by direct calculations. The additional conserved quantity $\mathbf{A}$ leads to the Coulomb degeneracy. This additional conservation means an additional symmetry, besides the symmetry under rotations.

Indeed, in equations (5.78) we may replace H by any discrete-level energy $-E$, and use instead of $\mathbf{A}$ the operator $\mathbf{u} = \mathbf{A}/\sqrt{-2E}$; equations (5.78) become

$$[l_i, u_j] = i\varepsilon_{ijk}u_k \ , \ [u_i, u_j] = i\varepsilon_{ijk}l_k \ ; \tag{5.79}$$

it is easy to see that $\mathbf{l}$ and $\mathbf{u}$ are the operators of infinitesimal rotations in a four-dimensional space; this is the additional Coulomb symmetry.[3] We introduce $\mathbf{j}_{1,2} = \frac{1}{2}(\mathbf{l} \pm \mathbf{u})$, which are independent: $[j_{\alpha i}, j_{\beta j}] = i\varepsilon_{ijk}j_{\alpha k}\delta_{\alpha\beta}$. In addition, $\mathbf{lu} = \mathbf{ul} = 0$ and $l^2 + u^2 = 4j_1^2 = 4j_2^2 = -1 - 1/2E$, or $4j^2 = -1 - 1/2E$, $E = -1/2[1 + 4j(j+1)] = -1/2(2j+1)^2 = 1/2n^2$, where n is any positive integer; the degeneracy is $(2j_1+1)(2j_2+1) = (2j+1)^2 = n^2$; since $\mathbf{l} = \mathbf{j}_1 + \mathbf{j}_2$, the orbital momentum l takes values up to $2j = n-1$. This is the Pauli matricial solution for the hydrogen atom.[4]

A similar extended symmetry occurs for the isotropic spatial oscillator with the hamiltonian $p^2/2m + m\omega^2r^2/2$, due to the squares of both p and r; with $\mathbf{a} = (m\omega\mathbf{r} + i\mathbf{p})/\sqrt{2m\hbar\omega}$ the hamiltonian becomes $\hbar\omega(\mathbf{a}^+\mathbf{a} + /2)$; this hamiltonian is invariant under any unitary group, which is wider than the group of the three-dimensional rotations. The additional quantum-mechanical symmetries for the Coulomb potential and the spatial oscillator are related to the closed orbits of the corresponding classical motions (which exist only for these potentials).

5.11 Concluding remarks

The ground-state wavefunction of the hydrogen atom is $\psi = e^{-r}/\sqrt{\pi}$ (in atomic units); the electron charge density is $\rho = -qe^{-2r}/\pi$, where $-q$ is the electron charge; the electric potential φ is given by $\Delta\varphi = -4\pi\rho$, or $\frac{1}{r}\frac{d^2}{dr^2}(r\varphi) = 4qe^{-2r}$; the solution of this equation is $\varphi = -q/r + q(1 + 1/r)e^{-2r}$; adding the proton contribution, we get the total potential $\varphi_t = q(1 + 1/r)e^{-2r}$. We can see that the electron screens off the proton potential q/r. There exists a finite number of bound states in this potential, which may explain the electron affinity.

[3] V. A. Fock, "Zur Theorie des Wasserstoffatoms", Z. Phys. **98** 145 (1935).

[4] W. Pauli, "Uber das Wasserstoffspektrum vom Standpunkt der neuen Quantenmechanik", Z. Phys. **36** 336 (1926).

Bound states exist also for potentials $V(r) = A/r^2 - B/r$ or $V(r) = A/r^2 + Br^2$ (with substitution $\xi \sim r^2$), where A, B are constants.

6 Perturbation Theory

6.1 Perturbations independent of time

Let $H_0\psi_n^{(0)} = E_n^{(0)}\psi_n^{(0)}$ be the equation for eigenvalues and eigenfunctions for the discrete spectrum of the hamiltonain H_0; a small perturbation V to H_0 brings small corrections to the eigenvalues and the eigenfunctions; we seek the wavefunction of $H_0 + V$ for the eigenvalue E as

$$\psi = \sum_{m=0} c_m \psi_m^{(0)} \quad ; \tag{6.1}$$

the equation $(H_0 + V)\psi = E\psi$ leads to

$$(E - E_k^{(0)})c_k = \sum_{m=0} V_{km} c_m \ . \tag{6.2}$$

We solve this equation by assuming

$$\begin{aligned} E &= E^{(0)} + E^{(1)} + E^{(2)} + \dots \ , \\ c_m &= c_m^{(0)} + c_m^{(1)} + c_m^{(2)} + \dots \ , \end{aligned} \tag{6.3}$$

where the superscripts 1 indicate quantities of the same order of smallness as V, the superscripts 2 indicate the second order, etc. For $c_n^{(0)} = 1$ and the rest zero we get

$$E^{(1)} = V_{nn} \ , \ E^{(2)} = {\sum_m}' \frac{| V_{mn} |^2}{E_n^{(0)} - E_m^{(0)}} \tag{6.4}$$

and

$$\psi_n = \psi_n^{(0)} + {\sum_m}' \frac{V_{mn}}{E_n^{(0)} - E_m^{(0)}} \psi_m^{(0)} + \dots \ , \tag{6.5}$$

where the prime means $m \neq n$. We can see that $\mid \psi_n \mid^1 = \mid \psi_n^{(0)} \mid^2$ up to the second order. The matrix elements of f are given by

$$f_{nm} = f_{nm}^{(0)} + \sum_k{}' \frac{V_{nk} f_{km}^{(0)}}{E_n^{(0)} - E_k^{(0)}} + \sum_k{}' \frac{f_{nk}^{(0)} V_{km}}{E_m^{(0)} - E_k^{(0)}} + \dots \ . \tag{6.6}$$

If the continuous spectrum is included

$$\psi_n = \psi_n^{(0)} + \sum_m{}' \frac{V_{mn}}{E_n^{(0)} - E_m^{(0)}} \psi_m^{(0)} + \int d\nu \frac{V_{\nu n}}{E_n^{(0)} - E_\nu} \psi_\nu^{(0)} \dots \ . \tag{6.7}$$

The results are valid for $\mid V_{mn} \mid \ll \mid E_n^{(0)} - E_m^{(0)} \mid$.

For a degenerate level E_n, with wavefunctions labelled by $n, n', \dots$ equation (6.2) yields

$$E^{(1)} c_n^{(0)} = \sum_{n'} V_{nn'} c_{n'}^{(0)} \ , \tag{6.8}$$

or

$$\sum_{n'} (V_{nn'} - E^{(1)} \delta_{nn'}) c_{n'}^{(0)} = 0 \ ; \tag{6.9}$$

the determinant of these equations should be equal to zero; this is called the secular equation. The perturbation removes, in general, the degeneracy. If $V_{nn'}$ are zero, or very small, we retain $c_m^{(1)}$ in equation (6.2), which yields now

$$E^{(1)} c_n^{(0)} = \sum_m V_{nm} c_m^{(1)} + \sum_{n'} V_{nn'} c_{n'}^{(0)} \ , \tag{6.10}$$

and

$$(E_n^{(0)} - E_m^{(0)}) c_m^{(1)} = \sum_{n'} V_{mn'} c_{n'}^{(0)} \ ; \tag{6.11}$$

we get $c_m^{(1)}$ from equation (6.11) and introduce it into equation (6.10); which leads to the secular equation with the change

$$V_{nn'} \rightarrow V_{nn'} + \sum_m \frac{V_{nm} V_{mn'}}{E_n^{(0)} - E_m^{(0)}} \ . \tag{6.12}$$

As an application, we may mention that the laplacian in elliptical coordinates brings perturbations which correspond to a deformed spatial potential well.[1]

6.2 Time-dependent perturbations

Let us assume that the perturbation $V(t)$ depends on time. We can solve the equation $[H_0 + V(t)]\psi = E\psi$ (perturbations included) and get time-dependent eigenvalues $E_n(t)$ and eigenfunctions $\psi_n(t)$; these eigenfunctions do not satisfy the time-dependent Schrodinger equation $i\hbar\partial\psi/\partial t = [H_0 + V(t)]\psi$, because $E_n(t)$ depend on time; these are not stationary states. We need a superposition of $\psi_n(t)$ with time-dependent coefficients (this is called the method of time-variation of the constants). Therefore, we seek the solution as

$$\psi = \sum_k a_k(t)\psi_k^{(0)} e^{-\frac{i}{\hbar}E_k^{(0)}t} \quad , \tag{6.13}$$

where $H_0\psi_k^{(0)} = E_k^{(0)}\psi_k^{(0)}$; we get

$$i\hbar\frac{da_m}{dt} = \sum_k V_{mk}(t)e^{\frac{i}{\hbar}(E_m^{(0)}-E_k^{(0)})t}a_k \; ; \tag{6.14}$$

it is convenient to introduce the notation $\hbar\omega_{mk} = E_m^{(0)} - E_k^{(0)}$.

We assume that the perturbation is small and $a_k = a_k^{(0)} + a_k^{(1)} + ...$, where $a_n^{(0)} = 1$, $a_k^{(0)} = 0$ for $k \neq n$; then, equation (6.14) becomes

$$i\hbar\frac{da_{kn}^{(1)}}{dt} = \sum_k V_{kn}(t)e^{i\omega_{kn}t} \quad , \tag{6.15}$$

whence

$$a_{kn}^{(1)} = -\frac{i}{\hbar}\int dtV_{kn}(t)e^{i\omega_{kn}t} \tag{6.16}$$

[1] L. Landau and E. Lifshitz, *Course of Theoretical Physics*, vol. 3, *Quantum Mechanics*, Elsevier (1977).

(including $k = n$); the wavefunction becomes

$$\psi_n = \psi_n^{(0)} e^{-\frac{i}{\hbar}E_n^{(0)}t} + \sum_k a_{kn}^{(1)} \psi_k^{(0)} e^{-\frac{i}{\hbar}E_k^{(0)}t} + ... ; \tag{6.17}$$

we can check the normalization (for $\omega_{kn} \neq \pm\omega$). If

$$V(t) = Fe^{-i\omega t} + Ge^{i\omega t} \tag{6.18}$$

with $F^+ = G$ (periodic perturbation), then

$$V_{kn}(t)e^{i\omega_{kn}t} = F_{kn}e^{i(\omega_{kn}-\omega)t} + F_{nk}^* e^{i(\omega_{kn}+\omega)} \tag{6.19}$$

and

$$a_{kn}^{(1)} = -\frac{F_{kn}e^{i(\omega_{kn}-\omega)t}}{\hbar(\omega_{kn}-\omega)} - \frac{F_{nk}^* e^{i(\omega_{kn}+\omega)t}}{\hbar(\omega_{kn}+\omega)} \tag{6.20}$$

(for $\omega_{kn} \neq \pm\omega$). The matrix elements of f are $f_{nm}(t) = f_{nm}^{(0)}(t) + f_{nm}^{(1)}(t) + ...$, where $f_{nm}^{(0)}(t) = f_{nm}e^{i\omega_{nm}t}$ and

$$\begin{aligned} f_{nm}^{(1)}(t) = &-e^{i\omega_{nm}t}\sum\nolimits_k \left[\frac{f_{nk}^{(0)}F_{kn}}{\hbar(\omega_{km}-\omega)} + \frac{F_{nk}f_{km}^{(0)}}{\hbar(\omega_{kn}+\omega)}\right] e^{-i\omega t} \\ &-e^{i\omega_{nm}t}\sum\nolimits_k \left[\frac{f_{nk}^{(0)}F_{mk}^*}{\hbar(\omega_{km}+\omega)} + \frac{F_{nk}^* f_{km}^{(0)}}{\hbar(\omega_{kn}-\omega)}\right] e^{i\omega t} . \end{aligned} \tag{6.21}$$

The formulae may be extended to the continuous spectrum, in which case the condition $E_{min}^{(0)} - E_n^{(0)} > 0$ should be satisfied, where $E_{min}^{(0)}$ is the lowest energy of the continuous spectrum.

6.3 Transitions

Let us assume $\omega_{mn} = \omega + \varepsilon$, where ε is small for a perturbation given by equation (6.19); then equation (6.14) becomes (mainly)

$$\begin{aligned} i\hbar\frac{da_m}{dt} &= F_{mn}e^{i(\omega_{mn}-\omega)t}a_n = F_{mn}e^{i\varepsilon t}a_n , \\ i\hbar\frac{da_n}{dt} &= F_{mn}^* e^{-i(\omega_{mn}-\omega)t}a_n = F_{mn}^* e^{-i\varepsilon t}a_m ; \end{aligned} \tag{6.22}$$

this system of equations can be solved by the substitution $b_n = e^{i\varepsilon t}a_n$; in the limit $\varepsilon \to 0$, the solution is

$$\psi(t) = \cos\Omega t \cdot e^{-\frac{i}{\hbar}E_m t}\psi_m^{(0)} - \frac{iF^*_{mn}}{\mid F_{mn}\mid}\sin\Omega t \cdot e^{-\frac{i}{\hbar}E_n t}\psi_n^{(0)} \quad , \qquad (6.23)$$

where $\psi(t=0) = \psi_m^{(0)}$ and $\hbar\Omega = \mid F_{mn}\mid$. The probability of the state n is $\sin^2\Omega t$; the state n is occupied periodically with period $1/2\Omega$, there exist periodic transitions between the two states.

6.4 Adiabatic hypothesis

Let us assume that the perturbation is vanishing at $t \to -\infty$ and lasts a finite duration of time (or falls rapidly to zero at $t \to \infty$); the wavefunction, for a discrete level n, becomes

$$\psi = \sum_k a_{kn}(t)\psi_k^{(0)}e^{-\frac{i}{\hbar}E_k^{(0)}t} \quad , \qquad (6.24)$$

where

$$a_{kn}(t) = a_{kn}^{(1)}(t) = -\frac{i}{\hbar}\int_{-\infty}^{t} dtV_{kn}(t)e^{i\omega_{kn}t} \quad , \ k \neq n \qquad (6.25)$$

and

$$a_{nn}(t) = 1 + a_{nn}^{(1)}(t) = 1 - \frac{i}{\hbar}\int_{-\infty}^{t} dtV_{nn}(t) \ ; \qquad (6.26)$$

at $t \to \infty$ the wavefunction is

$$\psi(\infty) = \sum_k a_{kn}(\infty)\psi_k^{(0)}e^{-\frac{i}{\hbar}E_k^{(0)}t} \quad , \qquad (6.27)$$

where $t \to \infty$; it differs from the initial wavefunction $\psi_n^{(0)}e^{-\frac{i}{\hbar}E_n^{(0)}t}$ for $t \to -\infty$; there exist finite probabilities

$$\mid a_{kn}(\infty)\mid^2 = \frac{1}{\hbar^2}\left|\int_{-\infty}^{+\infty} dtV_{kn}(t)e^{i\omega_{kn}t}\right|^2 \ll 1 \qquad (6.28)$$

for the states $k \neq n$. If the perturbation remains finite for $t \to \infty$, we integrate by parts and get

$$a_{fi} = -\frac{i}{\hbar}\int_{-\infty}^{t} dt V_{fi}(t) e^{i\omega_{fi}t} = \\ = -\left.\frac{V_{fi}e^{i\omega_{fi}t}}{\hbar\omega_{fi}}\right|_{-\infty}^{t} + \int_{-\infty}^{t} dt \frac{\partial V_{fi}}{\partial t}\frac{e^{i\omega_{fi}t}}{\hbar\omega_{fi}} \ , \tag{6.29}$$

where we introduced the suffixes f and i for the final and initial states. The first term gives the first-order perturbation contribution to the change of the wavefunction under the action of the constant perturbation (equation (6.5)); the transition probability is given by

$$w_{fi} = \frac{1}{\hbar^2\omega_{fi}^2}\left|\int_{-\infty}^{\infty} dt \frac{\partial V_{fi}}{\partial t} e^{i\omega_{fi}t}\right|^2 \ . \tag{6.30}$$

If the final state is in the continuous spectrum the transition probability is given by

$$dw_{fi} = \frac{1}{\hbar^2}\left|\int_{-\infty}^{+\infty} dt V_{fi}(t) e^{i\omega_{fi}t}\right|^2 d\nu_f \ , \tag{6.31}$$

where $d\nu_f$ is the infinitesimal number of states in the range ν_f to $\nu_f + d\nu_f$.

If the perturbation varies little during time intervals of the order of $1/\omega_{fi}$, then the integrals in equations (6.29) and (6.30) are vanishing: when the perturbation is introduced adiabatically, any non-degenerate stationary state is preserved (there are not transitions). This is the adiabatic hypothesis. In the opposite limit, when the interaction is introduced abruptly, equation (6.30) yields

$$w_{fi} = \frac{\mid V_{fi} \mid^2}{\hbar^2\omega_{fi}^2} \ ; \tag{6.32}$$

in this case there exist transitions to all states. Indeed, the transition probability in this case is given by

$$w_{fi} = \left|\int dq \psi_f^* \psi_i^{(0)}\right|^2 \ , \tag{6.33}$$

where $H_0\psi_i^{(0)} = E_i^{(0)}\psi_i^{(0)}$ and $(H_0 + V)\psi_f = E_f\psi_f$; these equations give $(E_f - E_i^{(0)}) \int dq\psi_f^*\psi_i^{(0)} = \int dq\psi_f^* V\psi_i^{(0)}$; if the perturbation is small we may replace E_f by $E_f^{(0)}$, ψ_f by $\psi_f^{(0)}$ and get the equation (6.32).

As an application we can compute the transition probabilities for a linear oscillator under the action of an electric field; the wavefunctions of the oscillator are displaced by the field and the matrix elements can be computed by using the properties of the Hermite polynomials. Another application is a sudden jolt applied to an atom, which receives a momentum $\hbar\mathbf{q} = \hbar m\mathbf{v}$, from a projectile with mass m and velocity $\mathbf{v}$. The wavefunction becomes $\psi_i e^{-i\mathbf{q}\sum\mathbf{r}}$, where the summation extends over all the coordinates of the electrons in the atom; the transition probability is given by equation (6.33); this way we can compute the ionization probability of the hydrogen atom. Similarly, we may compute the probability of ionization of an atom which suffers a β-decay, or the ionization probability of an atom associated with the α-decay.[2]

6.5 Transitions to the continuous spectrum

If the initial state, at $t = 0$, corresponds to a discrete level n, then the main contribution to the transition amplitude a_{fi} to a final state in the continuous spectrum comes from the term with $\omega_{fi} - \omega$ in equation (6.19); we assume $\hbar\omega > E_{min} - E_i^{(0)}$, where E_{min} is the lowest energy of the continuous spectrum. Equations (6.19) and (6.25) lead to

$$a_{fi} = -\frac{i}{\hbar}\int_0^t dt F_{fi} e^{i(\omega_{fi}-\omega)t} = -F_{fi}\frac{e^{i(\omega_{fi}-\omega)t} - 1}{\hbar(\omega_{fi} - \omega)} \tag{6.34}$$

and

$$\mid a_{fi} \mid^2 = \mid F_{fi} \mid^2 \frac{4\sin^2[\frac{1}{2}(\omega_{fi} - \omega)t]}{\hbar^2((\omega_{fi} - \omega)^2} \tag{6.35}$$

or

$$\mid a_{fi} \mid^2 = \frac{2\pi t}{\hbar^2} \mid F_{fi} \mid^2 \delta(\omega_{fi} - \omega) \ , \tag{6.36}$$

[2] See Landau and Lifshitz quoted above.

since $\sin^2 \alpha t/\pi t\alpha^2 \to \delta(\alpha)$ for $t \to \infty$. Therefore, the probability of transition per unit time is

$$dw_{fi} = \frac{2\pi}{\hbar} \mid F_{fi} \mid^2 \delta(E_f - E_i - \hbar\omega) d\nu_f \; ; \tag{6.37}$$

it looks like a resonant transition. Equation (6.37) is called "Fermi's golden rule".

The same result is obtained by adiabatic switching on the interaction:

$$a_{fi} = -\frac{i}{\hbar}\int_{-\infty}^{t} dt F_{fi} e^{i(\omega_{fi}-\omega)t+\alpha t} = -F_{fi}\frac{e^{i(\omega_{fi}-\omega)t+\alpha t}}{\hbar(\omega_{fi}-\omega-i\alpha)} \tag{6.38}$$

and

$$\mid a_{fi} \mid^2 = \frac{1}{\hbar^2} \mid F_{fi} \mid^2 \frac{e^{2\alpha t}}{(\omega_{fi}-\omega)^2+\alpha^2} \; ; \tag{6.39}$$

since $\alpha/(x^2+\alpha^2) \to \pi\delta(x)$ for $\alpha \to 0$ we get equation (6.37) per unit time.

If the perturbation V is constant ($\omega = 0$), equation (6.37) gives the transition rate from an initial state in the continuous spectrum to a final state in the continuous spectrum:

$$dw_{fi} = \frac{2\pi}{\hbar} \mid V_{fi} \mid^2 \delta(E_f - E_i) d\nu_f \; ; \tag{6.40}$$

both states have the same energy. The wavefunction is changed as

$$\Psi_i = \left[\psi_i^{(0)} + \int V_{fi}\psi_f^{(0)} \frac{d\nu_f}{E_i - E_f + i0}\right] e^{-\frac{i}{\hbar}E_i t} \tag{6.41}$$

(the contribution of final states in the discrete spectrum can be included); second-order contribution gives

$$\Psi_i = \{\psi_i^{(0)} + \int \left[V_{fi} + \int \frac{V_{f\nu}V_{\nu i}}{E_i - E_\nu + i0} d\nu\right] \cdot \\ \cdot \frac{\psi_f^{(0)} d\nu_f}{E_i - E_f + i0}\} e^{-\frac{i}{\hbar}E_i t} \tag{6.42}$$

and the transition rate is

$$dw_{fi} = \frac{2\pi}{\hbar}\left| V_{fi} + \int \frac{V_{f\nu}V_{\nu i}}{E_i - E_\nu + i0} d\nu \right|^2 \delta(E_f - E_i) d\nu_f \; ; \tag{6.43}$$

ν labels intermediate states.

Finally, we note that equation (6.35) implies an uncertainty $\Delta E \simeq 2\pi\hbar/\Delta t = h/\Delta t$ in energy.

6.6 Born cross-section

Let us suppose a collision of a particle with mass m on a potential $V(\mathbf{r})$; before collision, Schrodinger's equation is $(\Delta+k^2)\psi^{(0)} = 0$, during collision (and after collision) Schrodinger's equation is $(\Delta+k^2)\psi = (2mV/\hbar^2)\psi$, where $k^2 = 2mE/\hbar^2$; we assume an elastic collision. If V is small, it will determine a small change $\psi^{(1)}$ in the wavefunction, which is given approximately, by $(\Delta + k^2)\psi^{(1)} = (2mV/\hbar^2)\psi^{(0)}$; the particular solution of this equation is

$$\begin{gathered}\psi^{(1)}(\mathbf{r}) = -\frac{m}{2\pi\hbar^2}\int d\mathbf{r}'\psi^{(0)}(\mathbf{r}')\frac{V(\mathbf{r}')e^{ik|\mathbf{r}-\mathbf{r}'|}}{|\mathbf{r}-\mathbf{r}'|} \simeq \\ \simeq -\frac{me^{ikr}}{2\pi\hbar^2 r}\int d\mathbf{r}'V(\mathbf{r}')e^{i(\mathbf{k}-\mathbf{k}')\mathbf{r}'} = f\frac{e^{ikr}}{r} \quad ,\end{gathered} \tag{6.44}$$

where $\psi^{(0)} = e^{i\mathbf{kr}}$ and $\mathbf{k}'$ is the wavevector along the $\mathbf{r}$-direction (after scattering); we have assumed that V has a small range. Equation (6.44) is a spherical wave with amplitude f given by equation (6.44); the probability for the particle to pass through the surface dS in unit time is $vdS \mid f \mid^2 /r^2$, the incident flow is v, where v is the particle velocity; it follows that the cross-section for the solid angle do is $\mid f \mid^2 do$; it is convenient to introduce the notation $\mathbf{q} = \mathbf{k} - \mathbf{k}'$, $q = 2k\sin\theta/2$, θ being the scattering angle. This is the Born cross-section.[3]

The result given above by equation (6.44) is valid for

$$ma^2 \mid V \mid /\hbar^2 = V/(\hbar^2/ma^2) \ll 1 \quad , \tag{6.45}$$

where a is the range of the potential V and $ka < 1$; *i.e.* for potentials much smaller than the localization energy of the particle. Very shallow potentials satisfy this condition. This is valid in three dimensions, not in two or one dimensions.

[3]M. Born, "Quantenmechanik der Stossvorgange", Z. Phys. **38** 803 (1926).

The incident wavefunction may be taken as e^{ikx} (propagating along the direction x) and the solution of the equation $(\Delta + k^2)\psi^{(1)} = (2mV/\hbar^2)e^{ikx}$ is of the form $\psi^{(1)} = fe^{ikx}$; for high energies ($ka \gg 1$) we get

$$\psi^{(1)} = fe^{ikx} \simeq -\frac{im}{\hbar^2 k}e^{ikx}\int dxV \quad , \tag{6.46}$$

which leads to the condition $\mid U \mid \ll (\hbar^2/ma^2)ka = \hbar v/a$, where v is the velocity of the particle; this is a more relaxed condition of validity.

In the Coulomb field $V = \alpha/r$ the validity condition given above should be written as

$$\alpha \ll \hbar v \quad , \tag{6.47}$$

valid for any r. Therefore, for high energies, the Coulomb field may be regarded as a perturbation (except for trajectories close to the x-axis, where the integral in equation (6.46) diverges logarithmically).

For high energies the amplitude f obeys the approximate equation $2ik\partial f/\partial x = 2mVf/\hbar^2$, with the solution

$$f \sim e^{ikx}e^{-\frac{i}{\hbar v}\int dxV} \quad ; \tag{6.48}$$

at large x the equation solved above is not valid anymore, because of the derivatives with respect to the coordinates; so we must limit ourselves to $x \ll ka^2$.

6.7 Unitary transformations

Let ψ, φ... denote quantum states. U is a unitary transformation if it preserves the scalar products, *i.e.* $(U\varphi, U\psi) = (\varphi.\psi)$. The probabilities are then preserved, so unitary transformations describe symmetries. Probabilities are also preserved by $(U\varphi, U\psi) = (\varphi, \psi)^*$, and U is then called an anti-unitary transformation, which is also associated with a symmetry. This is a Wigner's theorem.[4]

Obviously a unitary transformation is such that $UU^+ = U^+U = 1$, so it is representable by $U = e^{iS}$, where S is hermitian, $S^+ = S$, and, therefore, an observable, in principle.

[4]E. P. Wigner, *Gruppentheorie und ihre Anwendung auf die Quantenmechanik der Atomspektren*, Vieweg (1931).

A unitary transformation preserves the commutation relations, *i.e.* the laws of motion, so it is also a canonical transformation.

Besides symmetries, another main application of the unitary transformations is their removing of interaction. Indeed, let the hamiltonian be $H = H_0 + \varepsilon V$, with usual notations, ε being a small parameter; and let $\psi = e^{i\varepsilon S}\psi'$. Then $H'\psi' = e^{-i\varepsilon S}He^{i\varepsilon S}\psi' = H_0\psi' + O(\varepsilon^2)$, providing $i\,[H_0, S] = V$. The effect of the interaction is now transferred upon the state $\psi = e^{i\varepsilon S}\psi'$. The interaction may depend on the time, so S depends on the time, and Schrodinger's equation requires

$$\partial S_{nk}/\partial t + i\omega_{nk}S_{nk} + V_{nk}/\hbar = 0 \ , \tag{6.49}$$

where $\omega_{nk} = (E_n - E_k)/\hbar$, $E_{n,k}$ are the eigenvalues of H_0. It follows

$$S_{nk} = -(1/\hbar)\int^t dt' \cdot V_{nk}(t')e^{i\omega_{nk}t'} \cdot e^{-i\omega_{nk}t} \ , \tag{6.50}$$

which is the amplitude of transition from state ψ_k' into state ψ_n'. The unitary transformation $e^{i\varepsilon S}$ is then the S-matrix or the evolution operator. It can also be written as $e^{iH_0t/\hbar}e^{-iH(t-t_0)/\hbar}e^{-iH_0t_0}$ for $\varepsilon = 1$, and reflects the symmetry under time translation.

7 Spin

7.1 Spin

Under rotations with angle $\delta\boldsymbol{\alpha}$ the orbital part ψ of the wavefunctions (*i.e.*, that part which depends on the positions $\mathbf{r}_a$, where a labels the particles) changes as $\widetilde{\psi} = U\psi$, $U = e^{\frac{i}{\hbar}\delta\boldsymbol{\alpha}\mathbf{L}}$, where $\mathbf{L} = \sum_a \mathbf{r}_a \times \mathbf{p}_a$ is the orbital angular momentum; p_a is the momentum of the a-th particle and $\hbar$ is Planck's constant; usually, we leave aside the constant $\hbar$ by the substitution $\mathbf{L} \to \hbar\mathbf{L}$, so $U = e^{i\delta\boldsymbol{\alpha}\mathbf{L}}$, where $\delta\boldsymbol{\alpha}\mathbf{L}$ is a dimensionless phase. The invariance under rotations (the symmetry under rotations, as for an isotropic space) implies $H\widetilde{\psi} = E\widetilde{\psi}$, where $H\psi = E\psi$, H being the hamiltonian and E the energy; we get $U^{-1}HU = H$ (or $U^{+}HU = H$, since $\mathbf{L}$ is hermitian and U is unitary), or $[H, U] = 0$, *i.e.* $[H, \mathbf{L}] = 0$; this commutation relation shows that $\mathbf{L}$ is conserved. From the commutation relations of the components of $\mathbf{L}$, we infer that only one component is determined simultaneously with L^2, say L_z; the eigenfunctions ψ_{lm} are labelled by l and m, where $l(l+1)$ and m are the eigenvalues of L^2 and L_z, respectively; l is an integer or a half-integer and m goes by unity from $-l$ to l; we have $2l+1$ m-states. For a rotation about the z-axis the wavefunctions ψ_{lm} are unchanged (up to a phase factor), but for rotations about other axes the wavefunctions become linear combinations of ψ_{lm}; these combinations are given by the rotation matrices; we say that ψ_{lm} are an irreducible representation of the rotation group (the rotation operators U form a group), with dimension $2l + 1$; the representation is irreducible since there is no sub-set of ψ_{lm} which is transformed into itself under rotations. We note that all these properties derive solely from the commutation relations of the operators of the orbital angular momentum.

An assembly of particles may be viewed as a composite particle; its orbital angular momentum is its intrinsic angular momentum. By extension, elementary particles may have an intrinsic angular momentum. This intrinsic angular momentum is called spin. The spin was suggested by Uhlenbeck and Goudsmit for electrons in 1925,[1] introduced in Quantum Mechanics by Pauli in 1927[2] and deduced by Dirac from his relativistic equation for electrons in 1930.[3] The electron has the spin $s = 1/2$. The great majority of elementary particles have spin $s = 1/2$; a few have spin zero; higher spins are not known for elementary particles. Since s is finite, the spin is vanishing in the classical limit $\hbar \rightarrow 0$ (this is not so for the orbital angular momentum where l may go to infinity for $\hbar \rightarrow 0$, thus giving a finite classical angular momentum). The commutation relations of the spin operators are those of the angular momentum, $[s_i, s_j] = i\varepsilon_{ijk}s_k$; the eigenvalues of s^2 are $s(s+1)$ and the eigenvalues of s_z ($2s+1$ eigenvalues) go from $-s$ to s ($-s$, $-s+1$, ...$-s+2$, $s-1$, s). The sum $\mathbf{j} = \mathbf{l} + \mathbf{s}$ is the total angular momentum of a particle with orbital angular momentum $\mathbf{l}$ and spin $\mathbf{s}$.

Apart from position coordinates (and other variables) the wavefunctions depend on the spin variable σ, $s_z\psi(\sigma) = \sigma^{'}\psi(\sigma)$; $\psi(\sigma)$ must carry the label $\sigma^{'}$, $\psi_{\sigma^{'}}(\sigma)$, and $\psi_{\sigma^{'}}(\sigma) \sim \delta_{\sigma\sigma^{'}}$; it is the spin component of the wavefunction. The scalar products include summations over σ. It follows $\mathbf{s}\psi(\sigma) = \mathbf{s}_{\sigma\sigma^{'}}\psi(\sigma^{'})$, with summation over $\sigma^{'}$; the spin operators are matrices (as expected), given by the matrix elements of the orbital angular momentum. For $s = 1/2$ ($\sigma = \pm 1/2$), we write $\mathbf{s} = \frac{1}{2}\boldsymbol{\sigma}$, where

$$\sigma_x = \begin{pmatrix} 0 & 1 \\ 1 & 0 \end{pmatrix} , \sigma_y = \begin{pmatrix} 0 & -i \\ i & 0 \end{pmatrix} , \sigma_z = \begin{pmatrix} 1 & 0 \\ 0 & -1 \end{pmatrix} \tag{7.1}$$

are the Pauli matrices; they have the properties $\{\sigma_i, \sigma_j\} = 2\delta_{ij}$ and $(\boldsymbol{\sigma}\mathbf{a})(\boldsymbol{\sigma}\mathbf{b}) = \mathbf{ab} + i\boldsymbol{\sigma}\cdot\mathbf{a}\times\mathbf{b}$; any 2×2 matrix can be written as a linear combination of Pauli matrices and the matrix unity.

---

[1] G. E. Uhlenbeck and S. A. Goudsmit, "Ersetzung der Hypothese vom unmechanischen Zwang durch eine Forderung bezuglich des inneren Verhaltens jedes einzelnen Elektrons", Naturwissenschaften **13** 953 (1925).

[2] W. Pauli, "Zur Quantenmechanik des magnetischen Elektrons", Z. Phys. **43** 601 (1927).

[3] P. A. M. Dirac, "The quantum theory of the electron", Proc. Roy. Soc. London **A117** 610 (1928); Part II, **A118** 351 (1928).

7.2 Spinors

The wavefunction of spin zero, $\mathbf{s}\psi(0) = 0$, is a scalar under rotations. Let us denote $\psi^1 = \psi(1/2)$, $\psi^2 = \psi(-1/2)$ for spin 1/2. Under any rotation the column ψ made of ψ^1, ψ^2 changes as $\psi^{'} = U\psi$, where

$$U = \begin{pmatrix} a & b \\ c & d \end{pmatrix} ; \tag{7.2}$$

we can check that $\psi^1\varphi^2 - \psi^2\varphi^1$ becomes, under rotations, $(ad - bc)(\psi^1\varphi^2 - \psi^2\varphi^1)$; therefore, it is a scalar and $detU = 1$. On the other hand,

$$U^{-1} = \begin{pmatrix} d & -b \\ -c & a \end{pmatrix} \tag{7.3}$$

and the unitarity condition $\psi^{'*}\psi^{'} = \psi^* U^+ U\psi = \psi^*\psi$ implies $U^+U = 1$, *i.e.* $U^+ = U^{-1}$; hence, $a = d^*$, $b = -c^*$, and

$$U = \begin{pmatrix} a & b \\ -b^* & a^* \end{pmatrix} ; \tag{7.4}$$

the matrix U is determined by three independent parameters, as many as the rotation angles. The ψ which rotates by U given by equation above is called a spinor; U is called a binary transformation. We note that ψ^{1*} and ψ^{2*} transform as ψ^2 and $-\psi^1$, respectively.

We call the column spinor ψ the contravariant spinor, and introduce the covariant spinor with components

$$\psi_\lambda = g_{\lambda\mu}\psi^\mu \ , \ (g_{\lambda\mu}) = (g^{\lambda\mu}) = \begin{pmatrix} 0 & 1 \\ -1 & 0 \end{pmatrix} , \tag{7.5}$$

$g^\mu_\lambda = \delta^\mu_\lambda$; $g_{\lambda\mu}$ is the metric spinor (tensor); it offers the possibility to treat the spinors as tensors; we can raise or lower the labels, we can do contractions, we can multiply spinors to get higher-order spinors (ψ^λ is a spinor of rank one); the scalar product is $\psi_\lambda\varphi^\lambda$ (and $\psi_\lambda\psi^\lambda = 0$); $\psi^{\lambda*}$ are transformed under rotations as ψ_λ.

A spinor of rank n symmetrical with respect to all its indices is called a symmetrical spinor of rank n. It cannot be reduced by contractions to a spinor of a lower rank. Only a spinor of rank two can be antisymmetrical (since its indices take only two values); it is a scalar multiple of $g_{\mu\nu}$; for instance, $g_{\lambda\mu}\psi_\nu+g_{\mu\nu}\psi_\lambda+g_{\nu\lambda}\psi_\mu=0$, since it is antisymmetrical in all its three indices. In addition, we have $\psi_{\lambda\mu}\psi^\mu_\nu=-\psi_{\nu\mu}\psi^\mu_\lambda$, and, precisely, $\psi_{\lambda\mu}\psi^\mu_\nu=-\frac{1}{2}\psi_{\rho\sigma}\psi^{\rho\sigma}g_{\lambda\nu}$.

The tensorial product of n 1/2-spinors is related to the wavefunction of a particle with spin $s=n/2$; indeed, the component $\psi^{11\ldots}$, with n places for 1, corresponds to $\psi(1/2,1/2,...)$, and, under rotations, this component gives a symmetrical tensor of rank n; therefore, the wavefunction of the spin s is the symmetrical spinor of rank $n=2s$ (with $2s+1$ components). Obviously, the relation between the symmetrical spinor and the wavefunction $\psi(\sigma)$ for a spin s, where $\sigma=s,\ s-1,...-s$, is

$$\psi(\sigma)=\left[\frac{(2s)!}{(s+\sigma)!(s-\sigma)!}\right]^{1/2}\psi^{11\ldots22\ldots}\ , \tag{7.6}$$

where there exist $s+\sigma$ places for 1 and $s-\sigma$ places for 2; we can see that if we apply the operator s_z, which is a tensorial sum of s_z-operators for 1/2-spin, we get σ; the factor in equation (7.6) arises from the normalization

$$\sum_{\sigma=-s}^{s}\mid\psi(\sigma)\mid^2=\sum_{\lambda,\mu,\ldots=1}^{2}\mid\psi^{\lambda\mu\ldots}\mid^2\ ; \tag{7.7}$$

more generally,

$$\psi^{\lambda\mu\ldots}\varphi_{\lambda\mu\ldots}=\sum_\sigma(-1)^{s-\sigma}\psi(\sigma)\varphi(-\sigma)\ . \tag{7.8}$$

Making use of the matrix elements of the 1/2-spin operators, we get

$$\begin{aligned} s_x\psi^{11\ldots22\ldots}&=\tfrac{1}{2}(s+\sigma)\psi^{11\ldots22\ldots}+\tfrac{1}{2}(s-\sigma)\psi^{11\ldots22\ldots}\ ,\\ s_y\psi^{11\ldots22\ldots}&=-\tfrac{i}{2}(s+\sigma)\psi^{11\ldots22\ldots}+\tfrac{i}{2}(s-\sigma)\psi^{11\ldots22\ldots}\ ,\\ s_z\psi^{11\ldots22\ldots}&=\sigma\psi^{11\ldots22\ldots}\ ,\end{aligned} \tag{7.9}$$

where on the left there exist $s+\sigma$ places for 1 and $s-\sigma$ places for 2, as well as on the right of the third equation, while on the right of the first and second equations there are $s+\sigma-1$ places for 1 and

$s - \sigma - 1$ places for 2 in the first spinors and $s + \sigma + 1$ places for 1 and $s - \sigma - 1$ places for 2 in the latter spinors. By contracting two indices and symmetrizing the rest, any spinor of rank n reduces to symmetrical spinors of rank n, $n-2$, $n-4$, ... (the symmetrical spinors give the irreducible representations of the rotation group).

Obviously, these results apply also to eigenfunctions of the orbital angular momentum (and the total angular momentum). For instance,

$$\psi_{jm} = \left[\frac{(2j)!}{(j+m)!(j-m)!}\right]^{1/2} \psi_{11\ldots 22\ldots} \ , \tag{7.10}$$

where there are $j+m$ places for 1 and $j-m$ places for 2. For integral j the functions ψ_{jm} are the spherical harmonics.

The case $j=1$ corresponds to

$$\begin{aligned} Y_{10} &= i\sqrt{\tfrac{3}{4\pi}}\cos\theta = i\sqrt{\tfrac{3}{4\pi}}\cos\theta \ , \\ Y_{1\pm 1} &= \mp i\sqrt{\tfrac{3}{8\pi}}\sin\theta e^{\pm i\varphi} = \mp i\sqrt{\tfrac{3}{8\pi}}(n_x \pm n_y) \ , \end{aligned} \tag{7.11}$$

where $\mathbf{n}$ is the unit vector along the radius vector. Making use of equation (7.10), we get

$$\begin{aligned} a_z &= i\sqrt{2}\psi^{12} \ , \\ a_x &= \tfrac{i}{\sqrt{2}}(\psi^{22} - \psi^{11}) \ , \ a_y = \tfrac{1}{\sqrt{2}}(\psi^{11} + \psi^{22}) \ , \end{aligned} \tag{7.12}$$

where $\mathbf{a}$ is any vector and $\psi^{\lambda\mu}$ is a symmetrical spinor of rank two ("vectors are equivalent with spin 1"); or

$$\mathbf{a} = \frac{i}{\sqrt{2}}\boldsymbol{\sigma}^{\lambda}_{\mu}\psi^{\mu}_{\lambda} \ , \ \psi^{\lambda}_{\mu} = -\frac{i}{\sqrt{2}}\mathbf{a}\boldsymbol{\sigma}^{\lambda}_{\mu} \ , \tag{7.13}$$

where $\boldsymbol{\sigma}^{\lambda}_{\mu}$ is a re-write of the matrix elements of the Pauli matrices. We can also check the formulae

$$\psi_{\lambda\mu}\varphi^{\lambda\mu} = \mathbf{ab} \ , \ \psi^{\lambda}_{\nu}\varphi^{\mu\nu} + \psi^{\mu}_{\nu}\varphi^{\lambda\nu} = \sqrt{2}\mathbf{a}\times\mathbf{b} \ . \tag{7.14}$$

In general, an irreducible tensor of rank (integral) j (with $2j+1$ components, like the spherical harmonics) and the symmetrical spinor of rank $2j$ give the same irreducible representation of the rotation group.

7.3 Finite rotations

The rotation matrix of angle α about the axis of the unit vector $\mathbf{n}$ for spin 1/2 is

$$U_{\mathbf{n}}(\alpha) = e^{i\frac{1}{2}\alpha\mathbf{n}\boldsymbol{\sigma}} = \cos\frac{\alpha}{2} + i\mathbf{n}\boldsymbol{\sigma}\sin\frac{\alpha}{2} \quad , \tag{7.15}$$

where $\boldsymbol{\sigma}$ are Pauli matrices; applied to the original spinor, it gives the rotated spinor; we note that for a rotation through angle 2π the spinors change sign. An arbitrary rotation is defined by Euler's angles α, β, γ: a rotation through angle α about the z-axis, a rotation through angle β about the new y-axis and a rotation through angle γ about the new z-axis (α and β are polar angles); it reads

$$U(\alpha,\beta,\gamma) = U_z(\gamma)U_y(\beta)U_z(\alpha) =$$

$$= \begin{pmatrix} e^{i(\alpha+\gamma)/2}\cos\frac{\beta}{2} & e^{-i(\alpha-\gamma)/2}\sin\frac{\beta}{2} \\ -e^{i(\alpha-\gamma)/2}\sin\frac{\beta}{2} & e^{-i(\alpha+\gamma)/2}\cos\frac{\beta}{2} \end{pmatrix} . \tag{7.16}$$

Let ψ_{jm} be the angular-momentum functions in the original frame and $\psi_{jm'}$ the angular momentum functions in the rotated frame; they are related through

$$\psi_{jm} = \sum_{m'} D^{(j)}_{m'm}(\alpha,\beta,\gamma)\psi_{jm'} \quad , \tag{7.17}$$

where $D^{(j)}$ is a (unitary) matrix of order $2j+1$; it is called the finite-rotation matrix. For spin 1/2 it is the matrix U given above; it can be written as '$e^{im'\gamma}d^{1/2}e^{im\alpha}$; by successive applications (as for symmetrical spinors), we get $e^{im'\gamma}d^{(j)}e^{im\alpha}$, where

$$d^{(j)}_{m'm} = \left[\frac{(j+m')!(j-m')!}{(j+m)!(j-m)!}\right]^{1/2} (\cos\beta/2)^{m'+m} .$$
$$\cdot(\sin\beta/2)^{m'-m} P^{(m'-m,m'+m)}_{j-m'}(\cos\beta) \quad , \tag{7.18}$$

$$P^{(a,b)}_n(z) = \frac{(-1)^n}{2^n n!}(1-z)^{-a}(1+z)^{-b}\frac{d^n}{dz^n}[(1-z)^{a+n}(1+z)^{b+n}] \tag{7.19}$$

being the Jacobi polynomials.[4] The rotation matrices have many symmetry properties; in particular they are orthogonal

$$\int \frac{d\omega}{8\pi^2} D^{(j_1)*'}_{m_1' m_1} D^{(j_2)}_{m_2' m_2} = \frac{1}{2j_1+1} \delta_{j_1 j_2} \delta_{m_1 m_2} \delta_{m_1' m_1'} \ , \tag{7.20}$$

where $d\omega = \sin\beta d\alpha d\beta d\gamma$ (they are an irreducible representation of the rotation group). For $j = 1$ we have

$$d^{(1)} = \begin{pmatrix} \frac{1}{2}(1+\cos\beta) & \frac{1}{\sqrt{2}}\sin\beta & \frac{1}{2}(1-\cos\beta) \\ -\frac{1}{\sqrt{2}}\sin\beta & \cos\beta & \frac{1}{\sqrt{2}}\sin\beta \\ \frac{1}{2}(1-\cos\beta) & -\frac{1}{\sqrt{2}}\sin\beta & \frac{1}{2}(1+\cos\beta) \end{pmatrix} \ , \tag{7.21}$$

where the first element corresponds to the pair $m'm = 11$. We have also $Y_{lm}(\beta, \alpha) = i^l \sqrt{(2l+1)/4\pi} D^{(l)}_{0m}(\alpha, \beta, \gamma)$.

7.4 Polarization of particles

For any particle with spin 1/2 there exists a direction along which $\sigma = 1/2$; the particle is completely polarized. An unpolarized particle is described by the density matrix (polarization matrix); it is a spinor $\rho^{\lambda\mu}$ with the properties

$$\rho^{\lambda}_{\lambda} = \rho^1_1 + \rho^2_2 = 1 \tag{7.22}$$

and $\rho^{\lambda*}_{\mu} = \rho^{\mu}_{\lambda}$. For a pure spin state (*i.e.*, not a mixed one), the polarization matrix reduces to

$$\rho^{\mu}_{\lambda} = \psi^{\mu} \psi^{*}_{\lambda} \ , \tag{7.23}$$

where ψ^{λ} is the spin wavefunction. The mean value of s_z is given by

$$\overline{s}_z = \frac{1}{2}(\rho^1_1 - \rho^2_2) \ , \tag{7.24}$$

hence $\rho^1_1 = 1/2 + \overline{s}_z$, $\rho^2_2 = -1/2 - \overline{s}_z$. Since $\overline{s}_{\pm} = \psi^{\lambda*} s_{\pm} \psi^{\lambda}$, making use of the matrices $s_{\pm}$, we get $\overline{s}_+ = \psi^{1*}\psi^2$, $\overline{s}_- = \psi^{2*}\psi^1$ for a pure

[4] A. R. Edmonds, *Angular Momentum in Quantum Mechanics*, Princeton University Press, Princeton (1957).

state; it follows $\overline{s}_+ = \rho_1^2$, $\overline{s}_- = \rho_2^1$ in a mixed state; combining these formulae with the one for s_z, we get

$$\rho_\mu^\lambda = \frac{1}{2}(\delta_\mu^\lambda + 2\sigma_\mu^\lambda \cdot \overline{\mathbf{s}}) \tag{7.25}$$

for the polarization matrix. The polarization is determined by the mean spin. For complete polarization one component of the mean spin is 1/2 and the other two are zero; for an unpolarized state all the components of the mean spin are zero; $\rho = 2(\overline{s}_x^2 + \overline{s}_y^2 + \overline{s}_z^2)^{1/2}$ is the degree of polarization.

For a particle with spin s the polarization matrix is given by $\rho_{\rho\sigma...}^{\lambda\mu...}$ of rank $4s$, with $2s$ symmetrical upper indices and $2s$ symmetrical lower indices, satisfying

$$\rho_{\lambda\mu...}^{\lambda\mu...} = 1 \ , \ \rho_{\rho\sigma...}^{\lambda\mu...*} = \rho_{\lambda\mu...}^{\rho\sigma...} \ ; \tag{7.26}$$

it has $4s(s+1)$ components.

7.5 Time reversal, Kramers theorem

In the absence of the spin the time-reversed wavefunction (solution of the Schrodinger equation) is $\psi^{rev} = \psi^*$. The same operation (complex conjugation) for a spin s, applied to the symmetrical spinor of rank $2s$, leads to $\psi^{\lambda\mu...*} = \psi_{\lambda\mu...}^{rev}$, *i.e.* the contravariant spinor becomes covariant; the covariant tensor corresponds to wavefunctions with $-\sigma$ (indeed, the angular momentum changes sign under time reversal); therefore, for the wavefunctions we have

$$\psi_{s-\sigma}^{rev} = (-1)^{s-\sigma}\psi_{s\sigma}^* \ ; \tag{7.27}$$

for two successive time-reversals, $\psi_{s\sigma} \rightarrow (-1)^{2s}\psi_{s\sigma}$: for half-integral spin the wavefunctions change sign.

Under time-reversal the electric field remains unchanged, but the magnetic field changes sign. In the presence of the electric field we must have time-reversal symmetry. For a non-degenerate energy level $\psi_{\lambda\mu...}^{rev} = C\psi_{\lambda\mu...} = \psi^{\lambda\mu...*}$, where C is a constant and $\psi^{\lambda\mu...}$ is a spinor of rank n. This equation reads also

$$\psi^{\lambda\mu...} = C^*\psi_{\lambda\mu...}^* \ ; \tag{7.28}$$

we multiply by $g_{\alpha\lambda}g_{\beta\mu} = (-1)^n g^{\lambda\alpha} g^{\mu\beta}$, and get

$$\psi_{\lambda\mu\ldots} = C^*(-1)^n \psi^{\lambda\mu\ldots *} = (-1)^n \mid C \mid^2 \psi_{\lambda\mu} \,. \qquad (7.29)$$

it follows $\mid C \mid^2 = (-1)^n$; n is the sum $2\sum_a s_a$, where s_a denotes the spin of the a-th particle (this sum may be different from the total spin S of the assembly of particles). It follows that n must be even, *i.e.* the spin sum $\sum_a s_a$ must be an integer; for an integer spin sum the electric field removes the energy degeneracy, for a half-integral sum the electric field leaves a two-fold degeneracy (complex conjugate spinors correspond to a different state with the same energy; this is called Kramers theorem; for even n the spinor is "real" and $C = 1$).[5]

[5] H. A. Kramers, "Theorie generale de la rotation paramagnetique dans les cristaux", Proc. Kon. Akad. Wet. (Amsterdam) **33** 959 (1930); E. Wigner, "Uber die Operation der Zeitumkehr in der Quantenmechanik", Nachr. Akad. Ges. Wiss. Gottingen **31** 546 (1932).

8 Identical particles

8.1 The principle of indiscernability

Quantum identical particles cannot be identified, distinguished from one another; this is so, because their properties are probabilistic, and the probabilities change in time. This is the "principle of indistinguishability" ("indiscernability") of identical quantum particles. It follows that the wavefunction $\psi(\xi_1, \xi_2)$ for two identical particles, where $\xi_{1,2}$ denote the coordinates and the spin variable, changes by a phase factor under the permutation of the two particles, $\psi(\xi_2, \xi_1) = e^{i\alpha}\psi(\xi_1, \xi_2)$, and $e^{2i\alpha} = 1$, $e^{i\alpha} = \pm 1$; the wavefunction is either symmetrical or antisymmetrical; the hamiltonian is invariant under particle permutations. It follows that the wavefunction of many identical particles is either totally symmetrical or totally antisymmetrical.

Electrons in atoms, as far as they are described by single-particle states, must occupy each a distinct state; there could not be two or more electrons in the same state, otherwise the matter would collapse. This is the exclusion principle of Pauli.[1] Therefore, the wavefunction of the electrons must be antisymmetrical; when two electrons occupy the same state the antisymmetrical wavefunction is zero. On this basis the Fermi-Dirac statistics was constructed.[2] Particles which obey the Fermi-Dirac statistics are called fermions. On the contrary, photons may occupy the same state in an undetermined number; therefore, their wavefunction must be symmetrical. On this basis the Bose-Einstein statistics was constructed.[3] Particles which obey the Bose-

[1] W. Pauli, "Uber den Zusammenhang des Abschlusses der Elektronengruppen im Atom mit der Komplexstruktur der Spektren", Z. Phys. **31** 765 (1925).

[2] E. Fermi, "Zur Quantelung des idealen einatomigen Gases", Z. Phys. **36** 902 (1926); P. A. M. Dirac, "On the theory of Quantum Mechanics", Proc. Roy. Soc. London **A112** 661 (1926).

[3] S. N. Bose, "Planck's Gesetz und Lichtquantenhypothese", Z. Phys. **26** 178

Einstein statistics are called bosons. What distinguishes the electrons from photons? Measurements of spectra and thermal properties have shown that all the particles (states) with integral spin are bosons and all particles (states) with half-integral spin are fermions. This is called the spin-statistics theorem. (Polarization of particles may be viewed as their spin). The fields associated to particles with integral spin obey second-order differential equations which lead to positive energies providing the wavefunctions are symmetrical; on the contrary, spinors obey Dirac-like first-order differential equations, for which the energy is positive only if the wavefunctions are antisymmetrical. All these provide the connection between spin and statistics.[4] All this is valid under the assumption that the field theories are invariant under the combined time reversal, change conjugation and spatial inversion (this is called the TCP symmetry).

If an assembly is composed of N independent identical particles with wavefunctions $\psi_{p_i}(\xi_i)$, $i = 1, 2, ...N$ the symmetrical wavefunction of the assembly is

$$\psi = \left(\frac{N_1!N_2!...}{N!}\right)^{1/2} \sum \psi_{p_1}(\xi_1)\psi_{p_2}(\xi_2)...\psi_{p_N}(\xi_N) \ , \qquad (8.1)$$

where N_i is the number of particles in the state p_i ($\sum_i N_i = N$) and the summation is performed over all permutations P; for antisymmetrical wavefunction the signature $(-1)^P$ must be included in the above summation, which makes the wavefunction equal to the determinant of the single-particle wavefunctions (it is called the Slater determinant[5]); the pre-factor in front of the summation in equation (8.1) normalizes the wavefunction. For two particles we have

$$\psi_s = \frac{1}{\sqrt{2}} \left[\psi_{p_1}(\xi_1)\psi_{p_2}(\xi_2) + \psi_{p_1}(\xi_2)\psi_{p_2}(\xi_1)\right] \qquad (8.2)$$

(1924); "Warmegleichgewicht im Strahlungsfeld bei Anwesenheit von Materie", **27** 384 (1924); A. Einstein, "Quantentheorie des einatomigen idealen Gases ", Sitz. Berichte Preuss. Akademie Berlin 261 (1924); 3 (1925); 18 (1925).

[4]M. Fierz, "Uber die Relativistische Theorie kraftefreier Teilchen mit beliebigem Spin", Helv. Phys. Acta **12** 3 (1939); W. Pauli, "The connection between spin and statistics", Phys. Rev. **58** 716 (1940).

[5]J. C. Slater, "The theory of complex spectra", Phys. Rev. **34** 1293 (1929).

for the symmetrical wavefunction and

$$\psi_a = \frac{1}{\sqrt{2}} \left[\psi_{p_1}(\xi_1)\psi_{p_2}(\xi_2) - \psi_{p_1}(\xi_2)\psi_{p_2}(\xi_1) \right] \tag{8.3}$$

for the antisymmetrical one.

8.2 Exchange energy

Let us consider two identical particles with spin zero. If their hamiltonian does not depend on spin (*e.g.*, non-relativistic electric interactions do not), the wavefunction is a product of the spin wavefunction and the coordinate wavefunction; the wavefunction depending on coordinates is called orbital wavefunction; in our case, of zero-spin particles, the spin wavefunction is an irrelevant scalar. The orbital wavefunction must be symmetrical in this case (since the total spin is zero); it follows that all energy levels have symmetrical wavefunctions with even orbital momentum l (of the relative motion; since the orbital wavefunctions have parity $(-1)^l$).

The situation is different for two identical particles with spin 1/2 (*e.g.*, electrons). The spin wavefunctions of the two particles are either symmetrical (spin 1) or antisymmetrical (spin 0); therefore, the orbital wavefunction is antisymmetrical in the former case and symmetrical in the latter (since the total wavefunctions must be antisymmetrical). It follows that the energy levels corresponding to the two orbitals with different symmetries are distinct (if an energy level is degenerate, linear combinations of its wavefunctions can be endowed with the necessary symmetry). Therefore, the energy levels of the assembly depend on spin, even if the interaction does not. This difference in energy is called exchange energy (because it is related to the symmetry under permutation - exchange - of the two particles).[6]

Let us consider two identical particles each with spin s; the spin wavefunction of the assembly is a spinor of rank $4s$, with two groups of $2s$ labels, the spinor being symmetrical with respect to the labels in

[6] W. Heisenberg, "Mehrkorperproblem und Resonanz in der Quantenmechanik", Z. Phys. **38** 411 (1926); "Uber die Spektra von Atomsystemen mit zwei Elektronen", Z. Phys. **39** 499 (1926).

each group. In order to obtain the spinor corresponding to spin S we must perform $2s - S$ contractions, with one label from one group and the other label from the other group; and then symmetrize within each group; we get a symmetrical spinor of rank $2S$. On contraction, we get antisymmetrical quantities, therefore, when particles are inter-changed, we get a factor $(-1)^{2s-S}$. On the other hand, the total wavefunction acquires a factor $(-1)^{2s}$ on inter-changing the particles; therefore, the orbital part of the wavefunction acquires the factor $(-1)^S$ on this operation. It follows that the orbital wavefunction is symmetrical when the total spin is even and antisymmetrical when the total spin is odd. Therefore, the energy levels of this assembly depend on the total spin, the difference being the exchange energy. For integral s there are $(2s+1)(s+1)$ states with even S (and $s(2s+1)$ states with odd S); for half-integral s there are $s(2s+1)$ states with even S and $(2+1)(s+1)$ states with odd S.

Let us suppose a spin-independent interaction V between two electrons and their orbitals

$$\psi_{s,a} = \frac{1}{\sqrt{2}} [\psi_1(r_1)\psi_2(r_2) \pm \psi_1(r_2)\psi_2(r_1)] \tag{8.4}$$

(corresponding to total spin $S = 0, 1$); if we treat V as perturbation we get the energy $E_{s,a} = A \pm J$, where

$$\begin{aligned} A &= \int dr_1 dr_2 V \mid \psi_1(r_1) \mid^2 \mid \psi_2(r_2) \mid^2 \ , \\ J &= \int dr_1 dr_2 V \psi_1^*(r_1)\psi_2^*(r_2)\psi_1(r_2)\psi_2(r_1) \ ; \end{aligned} \tag{8.5}$$

J is called exchange integral; these energy levels can be associated to the eigenvalues of the spin-exchange operator $-\frac{1}{2}J(1 + 4\mathbf{s}_1\mathbf{s}_2)$, where $\mathbf{s}_{1,2}$ are the spin operators of the two electrons. We note that J depends on the overlap of the two single-particle wavefunctions. The spin-exchange operator can be generalized to many-particle assemblies.

8.3 Symmetry under permutations

A hamiltonian which does not depend on spin variables for an arbitrary number of identical particles is invariant under any permutation of particles. Any permutation is a product of transpositions of two particles (and the signature of the permutation is -1 to the power equal to the number of transpositions in that permutation); the number of permutations is $N!$ and they form a group (with the multiplication operation). The hamiltonian discussed here is invariant under the permutation group. Therefore, its (orbital) wavefunctions are classified by the irreducible representations of the permutation group. Among them, there exist the totally symmetric orbital and the totally antisymmetric orbital. The irreducible representations of the permutation group are given by the so-called Young diagrams.[7] For a Young diagram we separate the N particles in N_1,N_2,...groups and arrange them in rows, below one another, with decreasing length, each particle in a cell in a row; the first row has N_1 cells, the next one has N_2 cells, etc. We symmetrize the wavefunction with respect to the particles in every row and antisymmetrize the wavefunction with respect to all particles in different rows (after antisymmetrization the symmetrization in each row is lost). For various distributions of particles in a Young diagram we get in this way the wavefunctions of an irreducible representation of the permutation group. Under any permutation the wavefunctions in any group transform into linear combinations of themselves (though not all are independent, those which are independent form the basis of the representation).

8.4 Second quantization

Let us assume N identical non-interacting bosons, each with single-particle ortho-normalized wavefunctions $\psi_i(\xi)$, where i labels the states and ξ denotes the coordinates (including the spin variable) for each particle; we leave aside the label of particles. Let us construct the operator

$$\psi(\xi) = \sum_i a_i \psi_i(\xi) \ , \tag{8.6}$$

where the operators a_i satisfy the commutation relations

$$[a_i, a_j] = 0 \ , \ [a_i, a_j^+] = \delta_{ij} \ . \tag{8.7}$$

[7]H. Weyl, *The Theory of Groups and Quantum Mechanics*, Methuen (1931); D. E. Rutherford, *Substitutional Analysis*, Edinburgh University Press (1948).

Obviously, the operators a and a^+ act in a space of states where there exist a vacuum $\mid 0>$ and states $\mid n>$ with any (integral) number of particles n, such that

$$a \mid 0>= 0 \ , \ a \mid n>= \sqrt{n} \mid n-1> \ ,$$
$$a^+ \mid n>= \sqrt{n+1} \mid n+1> \ , \ a^+a \mid n>= n \mid n> \ ; \tag{8.8}$$

since the quantization is applied to the operators a, a^+ this representation is called second quantization (the first quantization is related to the wavefunctions ψ_i).[8] For obvious reasons a is called annihilation (destruction, lowering) operator, $a^+$is called creation (raising) operator and a^+a is the operator of the number of particles in a state (occupation-number operators); the state $\mid n>$ is constructed by $\mid n>= (a^+)^n \mid 0> /\sqrt{n!}$; the space of states $\mid n>$ is called the Fock space, or the space of the particle numbers (we leave aside the label i of states).[9] The full occupation-number states are tensorial products of all individual states, of the form $\mid n_1 n_2, ... >$, which denotes n_1 particles in the state 1, n_2 particles in the state 2, etc. When applied to the vacuum, the operator $\psi(\xi)$ gives zero, while the operator $\psi^+(\xi)$ creates a superposition of particles, one in each state, all with the same coordinates ξ; we say that it creates a particle at point ξ (in an undetermined state), while $\psi(\xi)$ destroys a particle at point ξ (in an undetermined state); the destruction and creation operators $\psi(\xi)$, $\psi^+(\xi)$ are called field operators. They satisfy the commutation relations

$$[\psi(\xi), \psi(\xi^{'})] = 0 \ , \ [\psi(\xi), \psi^+(\xi^{'})] = \delta(\xi - \xi^{'}) \ . \tag{8.9}$$

The total number of particles is given by the operator

$$N = \int d\xi \psi^+(\xi)\psi(\xi) = \sum_i a_i^+ a_i \ . \tag{8.10}$$

Let us assume that we have a one-particle operator

$$f^{(1)} = \sum_a f_a^{(1)} \ , \tag{8.11}$$

[8]P. A. M. Dirac, "The quantum theory of emission and absorption of radiation", Proc. Roy. Soc. London **A114** 243 (1927).

[9]V. Fock, "Konfigurationsraum und zweite Quantelung", Z. Phys. **75** 622 (1932).

where $f_a^{(1)}$ acts on the coordinates of the a-particle; its elements of matrix tell that this operator destroys a particle in one state and creates a particle in another state; the construction

$$f^{(1)} = \int d\xi \psi^+(\xi) f^{(1)}(\xi)\psi(\xi) = \sum_{ij} f_{ij}^{(1)} a_i^+ a_j \tag{8.12}$$

does exacly the same; equation (8.12) gives the second-quantization representation of one-particle operators. In particular, for a one-particle hamiltonian $H^{(1)}$ with energies ε_i and wavefunctions ψ_i we have the representation

$$H^{(1)} = \sum_i \varepsilon_i a_i^+ a_i \tag{8.13}$$

A two-particle operator is given by

$$f^{(2)} = \sum_{a>b} f_{ab}^{(2)} \ , \tag{8.14}$$

where $f_{ab}^{(2)}$ acts on the coordinates of the two particles labelled by a and b; obviously,

$$\begin{aligned} f^{(2)} &= \tfrac{1}{2} \int d\xi_1 d\xi_2 \psi^+(\xi_1)\psi^+(\xi_2) f^{(2)}(\xi_1, \xi_2)\psi(\xi_2)\psi(\xi_1) = \\ &= \tfrac{1}{2} \sum_{ij} f_{ij,kl}^{(2)} a_i^+ a_j^+ a_k a_l \end{aligned} \tag{8.15}$$

is the second-quantization representation of this operator, where

$$f_{ij,kl}^{(2)} = \int d\xi_1 d\xi_2 \psi_i^*(\xi_1)\psi_j^*(\xi_2) f^{(2)}(\xi_1, \xi_2)\psi_k(\xi_2)\psi_l(\xi_1) \tag{8.16}$$

We can do the same second-quantization construction for fermions.[10] Let

$$\psi(\xi) = \sum_i c_i \psi_i(\xi) \tag{8.17}$$

[10] P. Jordan and E. Wigner, "Uber das Paulische Equivalenzverbot", Z. Phys. **47** 631 (1928).

be the field operator where the operators c_i satisfy the anticommutation relations

$$\{c_i, c_j\} = 0 \ , \ \{c_i, c_j^+\} = \delta_{ij} \ ; \qquad (8.18)$$

they act in a space of states where there exist a vacuum $\mid 0>$ and a state $\mid 1>$, such that

$$\begin{gathered} c \mid 0>= 0 \ , \ c \mid 1>=\mid 0> \ , \ c^+ \mid 0>=\mid 1> \ , \\ c^+ \mid 1>= 0 \ , \ c^+ c \mid 1>=\mid 1> \ ; \end{gathered} \qquad (8.19)$$

c is the annihilation (destruction) operator, c^+ is the creation operator and c^+c is the operator of the number of particles in a state (occupation-number operators). The full occupation-number states are tensorial products of all individual states, of the form $\mid n_1 n_2, ... >$, which denotes $n_1 = 0, 1$ particles in the state 1, $n_2 = 0, 1$ particles in the state 2, etc. The application of the operators c_i, c_i^+ on a state with many particles yields a factor $(-1)^p$, where p is the number of occupied states up to i; this factor arises from the anti-commutation relations (Jordan-Wigner factor). When applied to the vacuum the operator $\psi(\xi)$ gives zero, while the operator $\psi^+(\xi)$ creates a superposition of particles, one in each state, all with the same coordinates ξ; it creates a particle at point ξ (in an undetermined state), while $\psi(\xi)$ destroys a particle at point ξ (in an undetermined state). The field operators satisfy the anticommutation relations

$$\{\psi(\xi), \psi(\xi^{'})\} = 0 \ , \ \{\psi(\xi), \psi^+(\xi^{'})\} = \delta(\xi - \xi^{'}) \ . \qquad (8.20)$$

The total number of particles is given by the operator

$$N = \int d\xi \psi^+(\xi)\psi(\xi) = \sum_i c_i^+ c_i \ . \qquad (8.21)$$

and the one- and two-particle operators are given by formulae similar to the formulae given by equations (8.12) and (8.15).

Different kinds of particles may be viewed as different states of the same particles, so we may use the second-quantization formalism in this case, with additional labels for states.

8.5 Many-body theories

The second quantization aims at estimating the effect of the interaction starting with an assembly of free identical particles. The methods devised to this end are called many-body (many-particle) theories.[11] However, this goal raises a fundamental difficulty, because the wavefunction of an assembly of several particles is a multi-variable function $\psi(\xi_1,\xi_2,...)$, where $\xi_{1,2,...}$ are the coordinates of the particles, and the starting point of the theory is a wavefunction which is a superposition of products of single-particle wavefunctions of the (symmetrized) form $\psi_1(\xi_1)\psi_2(\xi_2)...$. The many-particle assemblies are entangled, and entangled wavefunctions cannot be constructed, rigourously, from wavefunctions of independent particles. Nevertheless, the many-body theories, though encountering serious difficulties, provide an interesting support for the more appropriate notion of elementary excitations (in particular, quasi-particles), introduced in the Theory of Condensed Matter by Landau.[12]

Let us suppose an assembly of N identical free particles enclosed in a finite volume Ω, sufficiently large to view the assembly as a macrocopic body. The one-particle wavefunctions are plane waves $e^{i\mathbf{kr}}/\sqrt{\Omega}$, multiplied by spin wavefunctions, where $\mathbf{k}$ is the wavevector and $\mathbf{r}$ denotes the position of the particle. We assume discrete values for wavevectors, by imposing cyclic boundary conditions on the boundaries of the (rectangular) volume Ω;[13] the Fourier transform of the interaction potential $V(\mathbf{r})$ (which may depend on spin) is

$$V(\mathbf{q}) = \int d\mathbf{r}V(\mathbf{r})e^{-i\mathbf{qr}} \ , \ V(\mathbf{r}) = \frac{1}{\Omega}\sum_{\mathbf{q}} V(\mathbf{q})e^{i\mathbf{qr}} \ . \qquad (8.22)$$

The hamiltonian of the assembly (equations (8.12) and (8.15)) can be

[11] See, for instance, D. J. Thouless, *The Quantum Mechanics of Many-Body Systems*, Academic Press (1961).

[12] L. Landau, "The theory of a Fermi liquid", Sov. Phys.-JETP **3** 920 (1957) (ZhETF **30** 1058 (1956)); "Oscillations in a Fermi liquid", Sov. Phys.-JETP **5** 101 (1057) (ZhETF **32** 59 (1957)); "On the theory of the Fermi liquid", Sov. Phys.-JETP **8** 70 (1959) (ZhETF **35** 97 (1958)).

[13] In macroscopic bodies short-range forces act, including screened Coulomb forces; matter with purely long-range (Coulomb) forces would collapse; the shape of the boundaries and boundary conditions for short-range forces are irrelevant.

written as

$$H=\sum_{\mathbf{k}}\varepsilon_{\mathbf{k}}a^{+}_{\mathbf{k}\alpha}a_{\mathbf{k}\alpha}+\frac{1}{2\Omega}\sum_{\mathbf{k}_1\mathbf{k}_2\mathbf{q}}V_{\alpha\beta}(\mathbf{q})a^{+}_{\mathbf{k}_1+\mathbf{q}\alpha}a^{+}_{\mathbf{k}_2-\mathbf{q}\beta}a_{\mathbf{k}_2\beta}a_{\mathbf{k}_1\alpha} \quad (8.23)$$

(summation of repeating spin indices is assumed); for bosons we use the letter a for the operators, for fermions we use the letter c, the form of the hamiltonian given by equation (8.23) being the same for the two types of particles; $\varepsilon_{\mathbf{k}}$ denotes the one-particle energies ($\hbar^2k^2/2m$, where m is the particle mass).

We may imagine that there exists a non-interacting ground-state of this assembly, with occupation numbers $n_{\mathbf{k}\alpha}=a^{+}_{\mathbf{k}\alpha}a_{\mathbf{k}\alpha}$ (without summation over α), and we would like to characterize the interacting ground-state and the excited states. The non-interacting ground-state for fermions is the Fermi sea, where all the **k**-states are occupied up to $k=k_F$, where the magnitude of the Fermi wavevector k_F is given by the fermion density $n=gk_F^3/6\pi^2$, g being the spin (s) multiplicity (*e.g.*, $g=2s+1$ if there does not exist a spin dependence). At this point, there appear difficulties. We envisage assemblies like electrons in metals, nucleons in atomic nuclei, $He^{3,4}$ atoms, photons, phonons, magnons, etc. First, for nucleons we neglect the finite-size effects (the atomic nuclei cannot be viewed as a macroscopic body); if the assembly is confined to an external potential, it should be included. For photons (or phonons) the number of particles is not defined (is not conserved). The most important difficulty arises from the fact that such assemblies are always at a finite temperature, as low as the temperature may be. In some cases the temperature effects can be neglected (as for nucleons), in other cases the temperature effects are small (like for electrons or extended nuclear matter), in some other cases the non-interacting ground-state as defined above does not exist; for instance, $He^{3,4}$ are superfluid at low temperature, electrons in metals become superconducting at low temperature, photons or phonons do not exist at low temperature. The Fermi sea is little distorted even at temperature as high as the room temperature (this is why the electrons are called degenerate).[14] If finite-temperature effects are to be considered, the quantum-mechanical knowledge consists of mean values of the physical quantities, given by means of the density matrix

[14]This is due to the high density of the electrons; we note that $1eV=1.16\times10^4K$.

for mixed states (statistical matrix); and only statistical quantities make sense at finite temperature, obtained by the statistical matrix. Therefore, our quantum-mechanical problem stated above becomes meaningless in these conditions. In condensed matter do not exist quantum states or energy levels.

Nevertheless, as a technical exercise, we may imagine that the interaction in equation (8.23) is small and the first-order contribution to the interacting ground-state is given by the mean value of the hamiltonian over the non-interacting ground-state; this type of calculation is called the Hartree-Fock approximation.[15] We can see that this calculation amounts to estimating the diagonal matrix elements of the interaction in equation (8.23), which occur for $\mathbf{q} = 0$ and $\mathbf{k}_2 = \mathbf{k}_1 + \mathbf{q}$. For bosons we get the Hartree-Fock energy

$$\begin{aligned} E_{HF} = \sum_{\mathbf{k}\alpha} \varepsilon_{\mathbf{k}} n_{\mathbf{k}\alpha} - \frac{1}{2\Omega} \sum_{\mathbf{k}} V_{\alpha\alpha}(\mathbf{q}=0) \left(n_{\mathbf{k}\alpha}^2 + n_{\mathbf{k}\alpha} \right) \\ + \frac{1}{2\Omega} \sum_{\mathbf{k}\mathbf{k}'} \left[V_{\alpha\beta}(\mathbf{q}=0) n_{\mathbf{k}\alpha} n_{\mathbf{k}'\beta} + V_{\alpha\alpha}(\mathbf{k}-\mathbf{k}') n_{\mathbf{k}\alpha} n_{\mathbf{k}'\alpha} \right] \; ; \end{aligned} \tag{8.24}$$

the terms with $\mathbf{q} = 0$ are Hartree terms, the term with $\mathbf{k} - \mathbf{k}'$ is the (exchange) Fock term. For fermions we get

$$\begin{aligned} E_{HF} = \sum_{\mathbf{k}\alpha} \varepsilon_{\mathbf{k}} n_{\mathbf{k}\alpha} - \frac{1}{2\Omega} \sum_{\mathbf{k}\mathbf{k}'} V_{\alpha\alpha}(\mathbf{k}-\mathbf{k}') n_{\mathbf{k}\alpha} n_{\mathbf{k}'\alpha} + \\ + \frac{1}{2\Omega} \sum_{\mathbf{k}\mathbf{k}'} V_{\alpha\beta}(\mathbf{q}=0) n_{\mathbf{k}\alpha} n_{\mathbf{k}'\beta} \; ; \end{aligned} \tag{8.25}$$

for electrons, the Coulomb repulsion given by the Hartree term $\mathbf{q} = 0$ in equation (8.25) is cancelled out by the Coulomb attraction between the electrons and the neutralizing rigid positive background and the Coulomb repulsion of the background; we are left with the exchange (Fock) term $\mathbf{k} - \mathbf{k}'$. The so-called "jellium" model is often used for electrons, where the positive neutralizing background is continuous (the results differ from those corresponding to a discrete, ionic background).

[15] D. R. Hartree, "The wave mechanics of an atom with a non-Coulomb central field. Part I. Theory and methods, Part II. Some results and discussion", Proc. Cambr. Phil. Soc. **24** 89, 111 (1928); V. Fock, "Naherungsmethode zur Losung des quantenmechanischen Mehrkorperproblems", Z. Phys. **61** 126 (1930).

By variation of the Hatree-Fock energy with respect to the occupation number $n_{\mathbf{k}\alpha}$ we get an energy associated to a particle; it has the form of a one-particle operator, which transformed back into the configuration space leads to the Hartree-Fock equations for single particles in an interacting assembly (a type of Schrodinger equation with an approximate interaction). This leads to the idea of one-particle Green functions, which are amplitudes of probability of particles propagating into an interacting assembly, and their generalization to multi-particle Green functions and response functions; Feynman diagrams are useful for computing them (as well as for computing the interacting ground-state energy).[16] These procedures give support to the quasi-particle elementary excitations, characterized by a finite lifetime.[17]

Another useful approximation in such type of computations is the so-called random phase approximation.[18] The particle density can be written as

$$\rho(\mathbf{r})=\sum_{a}\delta(\mathbf{r}-\mathbf{r}_a)=\frac{1}{\Omega}\sum_{\mathbf{q}}\left(\sum_{a}e^{-i\mathbf{r}_a}\right)e^{i\mathbf{qr}} \tag{8.26}$$

for each spin component; $\rho_{\mathbf{q}}=\sum_a e^{-i\mathbf{qr}_a}$ is a one-particle operator, which in the second quantization reads

$$\rho_{\mathbf{q}}=\sum_{\mathbf{k}}a^{+}_{\mathbf{k}}a_{\mathbf{k}+\mathbf{q}} \tag{8.27}$$

(for each spin component); the random phase approximation consists in keeping zero the sum $\mathbf{Q}$ of all $\mathbf{q}'$s which appear in calculation at various stages; it amounts to replacing $\sum_a e^{-i\mathbf{Qr}_a}$ by $\sum_a=N$, where N is the number of particles for a spin component; we can see that this approximation is based on the random distribution of the phases

[16] V. M. Galitskii and A. B. Migdal, "Application of quantum field theory methods to the many body problem", Sov. Phys.-JETP **7** 96 (1958) (ZhETF **34** 139 (1958)); A. Klein and R. Prange, "Perturbation theory for an infinite medium of fermions", Phys. Rev. **112** 994 (1958); P. C. Martin and J. Schwinger, "Theory of many-particle systems. I", Phys. Rev. **115** 1342 (1959).

[17] V. M. Galitskii, "The energy spectrum of a non-ideal Fermi gas", Sov. Phys.-JETP **7** 104 (1958) (ZHETF **34** 151 (1958)); "Sound excitations in Fermi systems", Sov. Phys.-JETP **7** 698 (1958) (ZhETF **34** 1011 (1058)).

[18] D. Bohm and D. Pines, "A collective description of electron interactions. III. Coulomb interactions in a degenerate electron gas", Phys. Rev. **92** 609 (1953).

in such types of summation, hence the name of the approximation. The motion of the operators $\rho_{\mathbf{q}}$ gives information about the density motion of the assembly, and, as such, it is related to the collective elementary excitations (distinct from quasi-particle excitations, but, similarly, characterized by a finite lifetime).

The structure of two-body forces of the hamiltonian has an important role in the Quantum Mechanics of many-particle assemblies. The superfluid ground-state is a Bose-Einstein condensate with a pair-wise structure,[19] the superconducting ground-state has a similar structure of Cooper pairs;[20] these states are entangled and their wavefunctions are coherent; the excitation states are elementary excitations given by one-particle hamiltonians, including collective modes. The two-body wavefunctions exhibit an interesting approach to the interaction in many-particle assemblies, like the so-called Bethe ansatz.[21]

Rigourosly speaking, the ground-state of interacting many-particle assemblies remains unknown; its low-lying elementary excitations govern the dynamics and determine the thermal properties. The quasi-particle notion introduced by Landau clarifies the relevance of the independent-electron model of Sommerfeld for electrons in metals; for fermions, there exist quasi-electrons above the Fermi surface, quasi-holes below the Fermi surface and the electron states lying just below the Fermi surface can be viewed as quasi-particles.

Finally, we note as a general remark that in the three-dimensional space we may exchange two identical particles by unconstrained rotations; this leads to fermions and bosons; in one dimension the rotations are impossible, which makes the fermions (electrons) to behave

[19] L. Landau, "The theory of superfluidity of He II", ZhETF **11** 592 (1941); "The theory of superfluidity of He II", J. Phys. (USSR) **5** 71 (1941); N. N. Bogoliubov, "On the theory of superfluidity", J. Phys. (USSR) **11** 23 (1947); N. N. Bogoliubov and N. Zubarev, Sov. Phys.-JETP **1** 83 (1955) (ZhETF **28** 129 (1955)).

[20] V. L. Ginsburg and L. Landau, "On the theory of superconductivity", ZhETF **20** 1064 (1950); J. Bardeen, L. N. Cooper and J. R. Schrieffer, "Microscopic theory of superconductivity", Phys. Rev. **106** 162 (1957); "Theory of superconductivity", **108** 1175 (1957).

[21] H. Bethe, "Zur Theorie der Metalle. I. Eigenwerte und Eigenfunktionen der Linearen Atomkette", Z. Phys. **71** 205 (1931); R. Jastrow, "Many-body problem with strong forces", Phys. Rev. **98** 1479 (1955).

as bosons;[22] this is the so-called Tomonaga-Luttinger model.[23] In two dimensions there exists a directional order, which allows rotations with angles which are fractions of 2π; this leads to a fractional statistics.[24]

[22]M. Apostol, "An electron crystal model in one dimension", Tr. J. Phys. **19** 1158 (1995).

[23]See, for instance, M. Apostol, "Jordan's boson representation for the one-dimensional two-fermion model", Phys. Lett. **A91** 177 (1982); "Bosonization of the one-dimensional two-fermion model", J. Phys. **C16** 5937 (1983) and references therein.

[24]R. B. Laughlin, "Anomalous quantum Hall effect: an incompressible quantum fluid with fractionally charged excitation" Phys. Rev. Lett. **50** 1395 9983).

9 Quasi-Classical Quantum Mechanics

9.1 Quasi-classical behaviour

In Classical Mechanics we measure position and momentum, or energy and time for the motion of a particle; it follows that we measure the mechanical action $S = -Et + px$, where x denotes position, t denotes the time, p is momentum and E is the energy. A more general situation would be to measure variations $\delta S = -E\delta t + p\delta x$ of the mechanical action; then, S would be the argument of a function, with a sufficiently general form $\psi(t, \mathbf{r}) = a(t, \mathbf{r})e^{\frac{i}{\hbar}S(t,\mathbf{r})}$, where $\mathbf{r}$ is the position vector, $\mathbf{p} = -gradS$ is the momentum vector and the Planck constant $\hbar$ is a mechanical action; the product of such functions would correspond to two or more particles, linear combinations of such functions would be a superposition of states (which requires a definition of the states), functions of several position variables would mean assemblies of many particles (distinct from products); all these are new situations, which remain to be given a meaning. A general form of mechanical motion may be written as $-dt/T + dx/\lambda = d\Phi$, where λ and T are some constants and Φ is some function of t and x; obviously, it suggests waves with wavelength λ and period T, of the form $\sim e^{i2\pi\Phi}$, where $\Phi = S/h$, $h = 2\pi\hbar$, according to what has been said above. Moreover, $E = h/T = h\nu = \hbar\omega$ and $p = h/\lambda = \hbar k$, where ω ($\nu = \omega/2\pi$) is frequency and $k = 2\pi/\lambda$ is the wavenumber (or the wavevector $\mathbf{k} = \mathbf{p}/\hbar$); these are the well-known quantization rules of Einstein for the energy-frequency and de Broglie for momentum-wavelength; $\hbar\omega$ is the energy quanta and $\hbar\mathbf{k}$ is the momentum quanta. We can see that such an enlargement of the mechanical framework leads to Quantum Mechanics and the wavefunction $\psi(t, \mathbf{r})$.

If the variation δS of the mechanical action is small, say, $\delta S \sim \hbar$, then the measurement process gives a comparable error $\Delta(\delta S) \simeq \delta S$, which implies

$$\Delta p \cdot \Delta x \simeq \hbar \ , \ \Delta E \cdot \Delta t \simeq \hbar \ ; \tag{9.1}$$

these are Heisenberg's uncertainty relations. We can see that the momentum can be measured with great accuracy only for large distances, and, similarly, the energy can be measured with great accuracy only for long times. The presence of a finite amount $\hbar$ of mechanical action, *i.e.* a mechanical-action quanta, which implies a quantum-mechanical behaviour, implies also an uncertainty in the measurement process (providing $\hbar$ is reasonably small). If the variation of the mechanical action is of order $n\hbar$, with an integer n large but $n\hbar$ still small, the uncertainty relations read $\Delta p \cdot \Delta x \gg \hbar$, $\Delta E \cdot \Delta t \gg \hbar$. If the errors are viewed as standard deviations, the largest errors for momentum and energy are $\Delta p \simeq p$ and $\Delta E \simeq E$, such that the minimal errors for position and time are $\Delta x \simeq \hbar/p = \lambda$ and $\Delta t \simeq \hbar/E = T$, in agreement with the wave-like nature of the motion.

On the contrary, if δS is very large, we may have $\Delta p \cdot \Delta x \ll \hbar$, $\Delta E \cdot \Delta t \ll \hbar$ which, in fact, means that the errors Δp, Δx, ΔE, Δt may be very small; in this situation, the most unfavourable case implies very small relative variations $\delta p/p \ll 1$, $\delta E/E \ll 1$. This limiting case is, in fact, the classical behaviour. Small variations with respect to the classical motion represent quasi-classical behaviour. That means that the external force F which acts upon the particle must be small in comparison with the force dp/dt generated by the uncertainties, *i.e.* $dp/dt \simeq pv/\lambda = p^3/\hbar m$:

$$\mid F \mid \ll \frac{p^3}{\hbar m} \ , \tag{9.2}$$

where m is the mass of the particle; since $p^3/\hbar m$ is of the order E/λ, it follows that the quasi-classical behaviour holds for $\mid gradV \mid \ll E/\lambda$, where V is the external potential which produces the force $\mathbf{F} = -gradV$. Equation (9.2) is equivalent with $d\lambda/dx \ll 1$, where $\lambda = \hbar/p$ is the wavelength and $p = \sqrt{2m(E-V)}$. Due to this circumstance, the quasi-classical behaviour is also known as the geometrical optics or ray approximation. The quasi-classical behaviour is defined by large variations of the mechanical action in comparison with $\hbar$; for separable motion, the mechanical action is the sum for all (separable) degrees of freedom.

The quasi-classical behaviour is defined by the mechanical action S (which has large variations); therefore, its wavefunction should be of the form $\psi = Ce^{\frac{i}{\hbar}S}$, where C is a constant. It follows

$$i\hbar\frac{\partial\psi}{\partial t} = E\psi \ , \ -i\hbar grad\psi = \mathbf{p}\psi \tag{9.3}$$

and

$$i\hbar\frac{\partial\psi}{\partial t} = -\frac{\hbar^2}{2m}\Delta\psi + V\psi = H\psi \ , \tag{9.4}$$

where H is the Hamilton function; therefore,

$$\psi = Ce^{-\frac{i}{\hbar}Et+\frac{i}{\hbar}\mathbf{pr}} \ , \ E = p^2/2m + V \ . \tag{9.5}$$

Equations (9.3) and (9.4), which follow from the existence of the wavefunction, have far-reaching implications. First, the existence of the wavefunction and the uncertainty relations mean that momentum and energy are not well-defined; they may take various distinct values. According to equations (9.3) and (9.4) these values can be obtained as eigenvalues for eigenfunctions of operators like $i\hbar\partial/\partial t$, $-i\hbar grad$, H. These distinct values of the mechanical quantities define quantum-mechanical states. Very conveniently, the eingenvalues of the mechanical quantities may be viewed as statistical variables, with mean values defined by bilinear forms of the wavefunction, of the form

$$\overline{O}(t) = \int d\mathbf{r}\psi^*(t,\mathbf{r})O(t,\mathbf{r})\psi(t,\mathbf{r}) \ , \tag{9.6}$$

where O is the operator associated to the physical quantity O; the complex conjugate of the wavefunction is necessary to have real mean values, such that the operators should be hermitian. It is easy to see that a superposition of eigenfunctions of an operator leads to a mean value which is a sum of weighted eigenvalues; it follows that the squared modulus of the amplitude of the eigenfunctions should be the probability of existence in that eigenstate. It follows that $\mid \psi(t,\mathbf{r}) \mid^2$ is the probability of localization of the particle in the volume element $d\mathbf{r}$, *i.e.* the number of particles in that volume element divided by the total number of particles; or the number of cases of finding the particle in the volume element $d\mathbf{r}$ in a measurement process, divided by the total number of measurements. We should have the normalization condition

$$\int d\mathbf{r} \left|\psi(t,\mathbf{r})\right|^2 = 1 \ , \tag{9.7}$$

where the integration is performed over the whole volume which contains the particle. Equation (9.4) implies a continuity equation for probabilities

$$\frac{\partial \mid \psi \mid^2}{\partial t} + div\mathbf{j} = 0 \ , \ \mathbf{j} = \frac{\hbar}{2mi}\left(\psi^* grad\psi - \psi grad\psi^*\right) \ , \tag{9.8}$$

where $\mid \psi \mid^2$ is the density of probability and $\mathbf{j}$ is the probability flow (flux), or the current density of probability, *i.e.* the number of particles per unit time and unit area perpendicular to $\mathbf{j}$ divided by the total number of particles; this is the probability conservation. For $\psi = Ce^{\frac{i}{\hbar}S}$ the current is $\mid C \mid^2 \mathbf{v}$, where $\mathbf{v} = \mathbf{p}/m$ is the velocity of the particle. For constant $\mathbf{v}$ the normalization condition given by equation (9.7) cannot be fulfilled for the whole space without special precautions, such that it is more convenient to normalize the wavefunction to the unit current, *i.e.* to take $C = 1/\sqrt{v}$. This is the statistical interpretation of the Quantum Mechanics due to Born.

The above description of the quantum-mechanical behaviour in the quasi-classical case is a consistent picture. It may be extended to all quantum-mechanical motions. Equation (9.4) is the Schrodinger equation; in its stationary form $H\psi = E\psi$, where the energy is well-defined and $\psi \sim e^{-\frac{i}{\hbar}Et}$, the energy levels E arise from a variational principle, as extrema of the mean hamiltonian. The statistical interpretation of the wavefunction can be extended to any type of states, the uncertainty relations are related to the non-commutability of the associated operators, the full determination of the quantum-mechanical behaviour is given by the set of all commuting physical quantities (complete description); the wavefunction can be extended to many-particle assemblies, where the wavefunctions are entangled; a behaviour which is only partly quantum-mechanical is described by the density matrix. The general wavefunction of the form $\psi = ae^{\frac{i}{\hbar}S}$

introduced into the Schrodinger equation leads to

$$\frac{\partial a^2}{\partial t} + \frac{1}{m} div(a^2 gradS) = 0 \ ,$$
$$\frac{\partial S}{\partial t} + \frac{1}{2m}(gradS)^2 + V - \frac{\hbar^2}{2ma}\Delta a = 0 \ , \tag{9.9}$$

which tells that in the limit $\hbar \rightarrow 0$ the classical description and the Hamilton-Jacobi equation are recovered, with S the mechanical action.

If we use the reduced mechanical action in $\psi = e^{\frac{i}{\hbar}S}$, then, from the stationary Schrodinger equation, we get

$$\frac{1}{2m}(gradS)^2 - \frac{i\hbar}{2m}\Delta S = E - V \ ; \tag{9.10}$$

hence, by $S = S_0 + \hbar S_1 + ...$, we get the quasi-classical wavefunction (for a coordinate)

$$\psi = \frac{C}{\sqrt{p}} e^{\frac{i}{\hbar}\int^x dx \cdot p} \ , \ p = \sqrt{2m(E-V)} \ , \tag{9.11}$$

where the normalization to the unit current appears by expansion in powers of $\hbar$. It is worth noting that the motion described by the wavefunction given by equation (9.11) is an almost free motion, where the change in the potential is compensated by changes in the momentum, such that the energy remains constant. It is worth noting that, in contrast with the classical motion, the particle is more localized in the regions where the potential is higher, due to the pre-factor $1/\sqrt{p}$; this unphysical feature shows that quantum-mechanical corrections to the quasi-classical (almost free) wavefunction, *i.e.* higher-order terms in powers of $\hbar$ (like the pre-factor $1/\sqrt{p}$) are meaningless; it is a substantial difference between p^2 which quasi-classically is $-\hbar^2(grad)^2$, and the quantum-mechanical p^2, which is $-\hbar^2\Delta$.

Equation (9.11) is valid for $E \gg V$. As long as $E < V$, or $E \simeq V$, *i.e.* in classically inaccessible regions of motion, where the motion is purely quantum-mechanical, the quasi-classical wavefunction given by equation (9.11) cannot be used; except for the case where it brings only small corrections. Such results are valid with "exponential accuracy". In these regions the momentum p becomes $\pm i \mid p \mid$

and the wavefunction decreases exponentially as $e^{\mp\frac{1}{\hbar}\int^x dx|p|}$, where $|p| = \sqrt{2m(V-E)}$. This describes a potential barrier, and the particle should have a large reflection coefficient by the potential barrier and a small penetration (tunneling) coefficient through the barrier, in accordance with the quasi-classical behaviour. The point given by $E = V(x)$ is a turning point. It follows that V should be large and wide inside the barrier, such that the wavefunction damping inside the barrier be large.

The quasi-classical wavefunction is also known as the JWKB approximation, after Jeffreys, Wentzel, Kramers and Brillouin.[1]

9.2 Wavefunction inside the barrier

Let us assume a turning point a, where $E = V(a)$, such that $E > V(x)$ for $x < a$ (outside the barrier) and $E < V(x)$ for $x > a$ (inside the barrier). Outside the barrier, the quasi-classical wavefunction is a superposition of incident and reflected travelling waves; inside the barrier the wavefunction is damped; we may write

$$\begin{gathered}\psi = \frac{C_1}{\sqrt{p}}e^{\frac{i}{\hbar}\int_a^x dx\cdot p} + \frac{C_2}{\sqrt{p}}e^{-\frac{i}{\hbar}\int_a^x dx\cdot p} \; , \; x < a \; , \\ \psi = \frac{C}{2\sqrt{|p|}}e^{-\frac{1}{\hbar}\int_a^x dx\cdot|p|} \; , \; x > a \; ,\end{gathered} \tag{9.12}$$

where $p = \sqrt{2m(E-V)}$ and $|p| = \sqrt{2m(V-E)}$. These wavefunctions are not valid for $x \simeq a$. For instance, we may expand the potential near $x = a$ and get $E - V = F(x-a)$, where $F = -\partial V/\partial x\,|_a$ is the force due to the barrier at a. This force should be small if the quasi-classical behaviour is valid, *i.e.* the potential should be shallow near a; on the other hand, $p = \sqrt{2mF(x-a)}$ should be large

[1] H. Jeffreys, "On certain approximate solutions of linear differential equations of the second order", Proc. London Math. Soc. **s2-23** 428 (1925); G. Wentzel, "Eine Verallgemeinerung der Quantenbedingungen fur die Zwecke der Wellenmechanik", Z. Phys. **38** 518 (1926); H. A. Kramers, "Wellenmechanik und Halbzahlige Quantisierung", Z. Phys. **39** 828 (1926); L. Brillouin, "La mecanique ondulatoire de Schrodinger: une methode generale de resolution par approximations successives", Comptes Rendus de l'Academie des Sciences **183** 24 ((1926).

in modulus, which implies a large deviation from $x = a$ where the quasi-classical behaviour is valid. However, we may assume a large F and integrate over x in the complex plane, circumventing the point $x = a$; this way, we get the exponential accuracy.

If we integrate from right to left with $x - a \gg e^{i\varphi}$, *i.e.* from $\varphi = 0$ to $\varphi = \pi$, the factor $p \sim \sqrt{x-a} \sim e^{i\varphi/2}$ yields $\cos(3\varphi/2) + i\sin(3\varphi/2)$ and $-i$ in the exponent for $\varphi = \pi$; we can see that the second row in equations (9.12) corresponds to the second term in the first row in equations (9.12); in addition, the pre-factor $1/2\sqrt{p}$ yields $e^{-i\pi/4}/2$; therefore, we should have $Ce^{-i\pi/4}/2 = C_2$. Similarly, if we integrate from 0 to $-\pi$ in the second row of equations (9.12) we get the first term in the first row of equations (9.12) and $Ce^{i\pi/4}/2 = C_1$. Therefore, we can write the wavefunction as

$$\psi = \tfrac{2C}{\sqrt{p}} e^{\frac{i}{\hbar}\int_a^x dx \cdot p + i\pi/4} + \tfrac{2C}{\sqrt{p}} e^{-\frac{i}{\hbar}\int_a^x dx \cdot p - i\pi/4} =$$

$$= \tfrac{C}{\sqrt{p}} \cos\left(\tfrac{1}{\hbar}\int_a^x dx \cdot p + \pi/4\right) \ , \ x < a \ , \tag{9.13}$$

$$\psi = \frac{C}{2\sqrt{|p|}} e^{-\frac{1}{\hbar}\int_a^x dx \cdot |p|} \ , \ x > a$$

and Kramer's correspondence rule

$$\frac{C}{2\sqrt{|\,p\,|}} e^{-\frac{1}{\hbar}\left|\int_a^x dx \cdot p\right|} \rightarrow \frac{C}{\sqrt{p}} \cos\left(\frac{1}{\hbar}\left|\int_a^x dx \cdot p\right| - \pi/4\right) \ , \tag{9.14}$$

irrespective of the side of the turning point on which the barrier lies. This relationship establishes a sort of "boundary conditions" for the quasi-classical wavefunction with exponential accuracy. For a potential barrier with an infinite vertical wall

$$\psi = \frac{C}{\sqrt{p}} \sin\left(\frac{1}{\hbar}\int_a^x dx \cdot p\right) \ , \ x < a \tag{9.15}$$

and $\psi = 0$ for $x > a$.

It is worth noting that Kramer's correspondence rule originates in the unphysical pre-factor $1/\sqrt{p}$, and circumventing the turning point through the complex plane does not provide physical boundary conditions. The rule given by equation (9.14) coincides with the result obtained by solving the Schrodinger equation for the potential

$V = E - F(x-a)$, which corresponds to a uniform electric field; but this is a quantum mechanical problem and the quasi-classical motion is not valid for a very shallow potential barrier. The quasi-classical wavefunction reduces in fact to

$$\psi = C_1 e^{\frac{i}{\hbar}\int_a^x dx\cdot p} + C_2 e^{-\frac{i}{\hbar}\int_a^x dx\cdot p} \; , \; x < a \; , \tag{9.16}$$

and

$$\psi = Ce^{-\frac{1}{\hbar}\int_a^x dx\cdot|p|} \; , \; x > a \; , \tag{9.17}$$

where $C_1 \simeq C_2$ and $\mid \psi \mid^2$ is very small for $x > a$, with exponential accuracy. These wavefunctions are continuous at the turning point for $C_1 + C_2 = C$ and the particle current is vanishing at the turning point, in accordance with the quasi-classical picture.

9.3 Bohr-Sommerfeld quantization rule

Let a particle move in a potential well $V(x)$, $b < x < a$. For high energy (and momentum) the quasi-classical wavefunction can be written in two distinct ways, according to equation (9.14):

$$\begin{aligned} \psi &= \frac{C}{\sqrt{p}}\cos\left(\frac{1}{\hbar}\int_x^a dx\cdot p - \pi/4\right) \; , \\ \psi &= \frac{C'}{\sqrt{p}}\cos\left(\frac{1}{\hbar}\int_b^x dx\cdot p - \pi/4\right) \; ; \end{aligned} \tag{9.18}$$

these two functions coincide only if

$$\frac{1}{\hbar}\int_b^a dx\cdot p - \pi/2 = \pi n \tag{9.19}$$

and $C' = C(-1)^n$, where n is a large, positive integer; equation (9.19) can also be written as

$$\begin{aligned} &\int_b^a dx\cdot p = \pi\hbar(n+1/2) \; , \\ &\frac{1}{2\pi}\oint dx\cdot p = \hbar(n+1/2) \; . \end{aligned} \tag{9.20}$$

These conditions show that the motion proceeds only on certain orbits, with discrete energies which follow from these equations by using

$p = \sqrt{2m[E - V(x)]}$; this is the Bohr-Sommerfeld quantization rule (also called the quantization of the "Old Quantum Mechanics"). The second row in equations (9.20) is the adiabatic invariant, which is constant for slow perturbations. The line integral in the second row of equations (9.20) shows that the surface enclosed by the curve $p(x)$ (in the phase space) is divided in cells with area $\Delta p \Delta x = 2\pi\hbar$; the integral number n defines the quantum-mechanical state; it is related to the number of oscillations of the wavefunction. The variation of the number of states in the second row of equations (9.20) gives

$$\delta E \oint dx \cdot \frac{\partial p}{\partial E} = \delta E \oint dx \cdot \frac{1}{v} = 2\pi\hbar \quad , \tag{9.21}$$

or $\delta E \simeq \hbar\omega$, where v is the velocity of the particle, T is the period of the motion and $\omega \simeq 2\pi/T$; the energy levels are equidistant, being represented approximately by $E = \hbar\omega n$ (where ω is not the same, in general, for any n). The pre-factor $1/\sqrt{p}$ in the wavefunction does not affect the quantization rule, except for the term 1/2 which should be replaced by a term γ of the order unity, depending on the boundary conditions (which are not included in the quasi-classical wavefunction).

The Bohr-Sommerfeld quantization rule originates in the period T of a periodic motion, given by

$$\oint dx/v = T \quad , \tag{9.22}$$

where v is the classical velocity and the integration is carried out over a closed orbit; in this equation we can write

$$\oint dx/v = \oint dx/(\Delta E/\Delta p) = \frac{1}{\Delta E}\Delta \oint dx \cdot p = T \quad , \tag{9.23}$$

for small variations of the adiabatic invariant, where p denotes the momentum. We note that $v = \partial E/\partial p$ is the (local) group velocity; the variation implied by equation (9.23) is in fact a quantum jump, which proceeds as the motion of a classical particle with (group) velocity and trajectory; also, equation (9.23) implies a dense distribution of energy levels and a superposition of states, which forms a wavepacket.

According to Bohr's correspondence principle,[2] we may associate this change with the Planck-Einstein quantum hypothesis $\Delta E = h\nu = h/T$, and get

$$\Delta \oint dx \cdot p = h \ ; \tag{9.24}$$

it follows that the adiabatic invariant changes by small, finite amounts h of mechanical action; since for slow perturbations Ehrenfest's adiabatic invariant[3] does not change (it has not a continuous change), it follows that it should be an integer of h; therefore, if the motion proceeds between a and b, we get

$$\int_a^b dx \cdot p = \frac{1}{2} hn = \pi \hbar n \ \ , \tag{9.25}$$

where n is a positive integer; this is the Bohr-Sommerfeld quantization rule given by equations (9.20). It holds for small variations from the classical behaviour, *i.e.* for large n, *i.e.* for the quasi-classical behaviour. The correspondence principle and the adiabatic invariants have been the guiding principles in formulating the Quantum Mechanics.[4]

In the quasi-classical regime the wavefunction is a superposition of eigenfunctions of energy, peaked on some n and extending over some Δn, $0 \ll \Delta n \ll n$ (a wavepacket); the mean value of some quantity f can be written as

$$\overline{f} = \sum_{mn} a_m^* a_n f_{mn} e^{i\omega_{mn}t} = \sum_{ns} a_{n+s}^* a_n f_{n+s,n} e^{i\omega st} \ \ , \tag{9.26}$$

where a_n are the coefficients of the superposition, $s = \Delta n$ and the energies are equidistant and separated by small quantities $\hbar\omega$; equation (9.26) can also be written approximately as

$$\overline{f} \simeq \sum_{ns} \mid a_n \mid^2 f_s e^{i\omega st} = f(t) \ \ , \tag{9.27}$$

[2]N. Bohr, "Uber die Serienspektra der Elemente", Z. Phys. **3** 423 (1920).

[3]P. Ehrenfest, "Adiabatic invariants and the theory of quanta", Phil. Mag. **33** 500 (1917); "On adiabatic changes of a system in connection with the quantum theory", Proc. Acad. Amst. **19** 576 (1917); "Adiabatische Invarianten und Quantentheorie", Ann. Phys. **51** 327 (1916).

[4]A. Sommerfeld, *Atombau und Spektrallinien*, Vieweg&Son, Braunschweig (1939).

where $f(t)$ is the classical quantity f; therefore, the matrix elements f_{mn} are the Fourier components f_{m-n} of the classical quantity f, as expected.

The quantization rule and the quasi-classical procedure can be applied to a finite, separable motion (which implies a complete separation of variables in the Hamilton-Jacobi equation; such a motion is called a conditionally periodic motion); the term 1/2 in the quantization rule given by equations (9.20) may be replaced in this case by a constant of the order of unity, which depends on the nature of the boundary conditions (*e.g.*, for a cyclic boundary condition this number is zero). For a general (finite) motion, the notion of cells with area $2\pi\hbar$ in the phase space is preserved, being related to the number of eigenoscillations (characteristic vibrations) of the wavefunction.

For a rectangular potential well with infinite walls at $x = 0$, a the wavefunctions are $(2/a)^{1/2}\sin \pi nx/a$ and the energy levels are $E_n = \pi^2\hbar^2 n^2/2ma^2$, where $n = 1, 2, ...$ The quantization rule of the quasi-classical motion (equations (9.20)) gives $\sqrt{2mE}a = \pi\hbar(n+\gamma)$, where γ is a constant of the order unity; it follows $E = E_n$ for $\gamma = 0$ and $p = \pi\hbar n/a$; the energy spectrum can be approximated by $\delta E = (\pi^2\hbar^2 n/ma^2)\delta n$ in the vicinity of n; we can see that the energy levels are equidistant for a limited range.

The number of states labelled by momentum is $(4\pi/3)p_{max}^3 dV/(2\pi\hbar)^3$; for a potential well U, such that $E = p^2/2m + U < 0$ (*i.e.* for quasi-classical states) $p_{max} = (-2mU)^{1/2}$ and the number of states is $(4\pi/3)\int(-2mU)^{3/2}dV/(2\pi\hbar)^3$; for $U \sim r^{-s}$, $s < 2$ this number of states is infinite. If the potential is centrally symmetric there exists an orbital degeneracy and the energy is given by $E = p_r^2/2m + U + L^2/2mr^2 < 0$; we have $p_{rmax} = [2m(-U - L^2/2mr^2)]^{1/2}$ and the number of energy levels $\int dp_r dr/(2\pi\hbar) = \int drp_{rmax}/(2\pi\hbar)$ for a given L; integrating with respect to $L/\hbar$ we get $(m/4\hbar^2)\int r(-U)dr$. For $s < 2$ there exists an infinite number of quasi-classical states lying far away.

9.4 Centrally-symmetric field

Let us consider an attractive potential $V(r) = -\alpha/r^s$, where the constant $\alpha > 0$. For $s > 2$ the particle falls to the centre (the origin), for $s = 2$ the motion depends on the angular momentum, but for $s < 2$ there exists an infinity of energy levels (bound states), denser towards small negative energies E. From

$$E = \frac{1}{2m}p^2 - \frac{\alpha}{r^2} \tag{9.28}$$

we can see that the momentum p is large for small r and is vanishing for large r. For such orbits, with small negative energies, the motion is quasi-classical. The kinetic energy T can be written as

$$2T = \mathbf{pv} = \frac{d}{dt}(\mathbf{pr}) - \mathbf{r}\frac{d\mathbf{p}}{dt} = \frac{d}{dt}(\mathbf{pr}) + \mathbf{r}\frac{dV}{d\mathbf{r}} \ , \tag{9.29}$$

where $\mathbf{v}$ is the velocity; we may integrate over a closed orbit (or an almost closed orbit) and get

$$2T = -sV \ , \tag{9.30}$$

since V is a homogeneous function. This is known as the virial theorem ($\mathbf{pr}$ is called the virial); we get

$$T = -\frac{s}{2-s}E \ , \ E = \frac{2-s}{2}V = -\frac{2-s}{2}\frac{\alpha}{a^s} \ , \tag{9.31}$$

where a is the mean radius of the orbit. The momentum becomes $p = \sqrt{[2ms/(2-s)] \mid E \mid}$ and the quantization rule gives

$$\sqrt{2ms \mid E \mid /(2-s)}a = n\hbar \ ; \tag{9.32}$$

inserting this result in equation (9.31) we get the energy levels

$$E_n = -\frac{2-s}{2}\alpha^{\frac{2}{2-s}}\left(\frac{sm}{\hbar^2 n^2}\right)^{\frac{s}{2-s}} \ ; \tag{9.33}$$

for the Coulomb potential ($s = 1$) they coincide with the exact quantum-mechanical result $E_n = -m\alpha^2/2\hbar^2 n^2$.

The velocity of a particle in spherical coordinates is

$$\mathbf{v} = (\dot{r},\, r\dot{\theta},\, r\sin\theta\dot{\varphi}) \tag{9.34}$$

and the lagrangian of a free particle with mass m reads

$$L = \frac{1}{2}m(\dot{r}^2 + r^2\dot{\theta}^2 + r^2\sin^2\theta\dot{\varphi}^2) \ ; \tag{9.35}$$

the momenta are

$$p_r = m\dot{r} \ , \ p_\theta = mr^2\dot{\theta} \ , \ p_\varphi = mr^2\sin^2\theta\dot{\varphi} \tag{9.36}$$

and the hamiltonian is

$$H = \frac{1}{2m}\left(p_r^2 + \frac{p_\theta^2}{r^2} + \frac{p_\varphi^2}{r^2\sin^2\theta}\right) \ . \tag{9.37}$$

If the potential V is of the form

$$V = v_1(r) + \frac{v_2(\theta)}{r^2} + \frac{v_3(\varphi)}{r^2\sin^2\theta} \tag{9.38}$$

the Hamilton-Jacobi equation $\partial S/\partial t + H(\partial S/\partial\mathbf{r}, \mathbf{r})$ $(E = H(\mathbf{p}, \mathbf{r}))$ is separable; we have

$$\begin{gathered} p_\varphi^2 + 2mv_3(\varphi) = \alpha \ , \ p_\theta^2 + \tfrac{\alpha}{\sin^2\theta} + 2mv_2(\theta) = \beta \ , \\ p_r^2 + \tfrac{\beta}{r^2} + 2mv_1(r) = 2mE \ , \end{gathered} \tag{9.39}$$

where α, β are constants and E is the energy (integrals of the motion); the momenta given by equations (9.39) can be tentatively used for the quasi-classical motion, but the orbits are not closed in general (they are closed only for the potentials $-r^2$and $-1/r$).

The problem is simplified for a central field, when V reduces to $V(r)$. In this case the angular momentum $\mathbf{L} = \mathbf{r} \times \mathbf{p}$ is a constant vector ($\dot{\mathbf{L}} = 0$). Consequently, the motion proceeds in a plane perpendicular to $\mathbf{L}$; in this plane we use polar coordinates r and φ; the velocities are $\dot{r}$ and $r\dot{\varphi}$ and the angular momentum $\mathbf{L}$ reduces to its component perpendicular to the plane which contains the trajectory, $L = mr^2\dot{\varphi}$. The lagrangian is

$$\mathcal{L} = \frac{1}{2}m(\dot{r}^2 + r^2\dot{\varphi}^2) - V(r) \ , \tag{9.40}$$

the momenta are

$$p_r = m\dot{r} \ , \ p_\varphi = mr^2\dot{\varphi} = L \tag{9.41}$$

and the hamiltonian is

$$H = \frac{1}{2m}p_r^2 + \frac{L^2}{2mr^2} + V \ . \tag{9.42}$$

The equations of motion

$$\tfrac{d}{dt}(mr^2\dot{\varphi}) = 0 \ , \ \tfrac{d}{dt}(m\dot{r}) = -\tfrac{\partial U}{\partial r} + \tfrac{L^2}{mr^3} \tag{9.43}$$

lead to the energy conservation

$$E = \frac{1}{2m}p_r^2 + \frac{L^2}{2mr^2} + U \ , \tag{9.44}$$

or

$$\dot{r} = \sqrt{\frac{2}{m}(E - V) - L^2/m^2r^2} \ ; \tag{9.45}$$

making use of $\dot{\varphi} = L/mr^2$ we get

$$\frac{d\varphi}{dr} = \frac{1/r^2}{\sqrt{\frac{2m}{M^2}(E - V) - 1/r^2}} \ , \tag{9.46}$$

which gives the parametric equation of the trajectory $\varphi(r)$.
The above equations show that φ is a monotonically increasing function of the time ($\dot{\varphi} = L/mr^2 > 0$); the coordinate r moves in the effective potential

$$V_{eff} = \frac{L^2}{2mr^2} + V \ , \tag{9.47}$$

which includes the centrifugal contribution $L^2/2mr^2$. The motion along the coordinate r is limited by the condition $E = U_{eff}(r)$; if there exists one solution r_{min} of this equation, the motion is infinite, $r > r_{min}$; if there exist two solutions r_{min} and r_{max} of this equation, the motion is finite, $r_{min} < r < r_{max}$. The points r_{min} and r_{max} are turning points, where $\dot{r} = 0$ (but, of course, $\dot{\varphi} \neq 0$). The change $\Delta\varphi$ in the angle for a motion from r_{min} forth to r_{max} and back to r_{min} is not a rational fraction of 2π, in general; then, trajectory is not closed. For $V \sim 1/r$ (Kepler's problem) and $V \sim r^2$ (space oscillator) the orbits are closed. Since

$$\frac{1}{2}m\dot{r}^2 = E - V - \frac{L^2}{2mr^2} > 0 \quad , \tag{9.48}$$

it follows that the particle may fall to the centre for

$$(r^2 U)_{r\to 0} < -L^2/2m \quad , \tag{9.49}$$

i.e. either $U \sim -\alpha/r^2$ with $\alpha > M^2/2m$, or $U \sim -1/r^\beta$ with $\beta > 2$. For the attractive Coulomb potential $V = -\alpha/r$ equation (9.46) can be integrated immediately by the change of variable $u = 1/r - m\alpha/L^2$; we get

$$p/r = 1 + e\cos\varphi \quad , \tag{9.50}$$

where

$$p = L^2/m\alpha \ , \quad e = \sqrt{1 + 2EL^2/m\alpha^2} \ ; \tag{9.51}$$

for $E < 0$ equation (9.50) describes an ellipse, with eccentricity e. For this closed trajectory we may use the quasi-classical procedure.

For a central field the mechanical action is

$$S = -Et + p_\varphi \varphi + \int^r drp_r \quad , \tag{9.52}$$

where $p_\varphi = L = const$; it follows that the adiabatic invariant for the coordinate φ is

$$I_\varphi = \frac{1}{2\pi}\int_0^{2\pi} d\varphi L = L \tag{9.53}$$

and the adiabatic invariant for the coordinate r

$$\begin{aligned} I_r &= \tfrac{1}{2\pi} 2\int_{r_{min}}^{r_{max}} drp_r = \\ &= \tfrac{1}{\pi}\int_{r_{min}}^{r_{max}} dr\sqrt{2m(-\mid E\mid - V) - L^2/r^2} \ , \end{aligned} \tag{9.54}$$

according to equations (9.39). Making use of $p_r = m\dot{r}$ and equation (9.50) for the Coulomb potential, we have

$$\begin{aligned} I_r &= \tfrac{m}{\pi}\int_\pi^0 d\varphi r'^2\dot{\varphi} = \tfrac{L}{\pi}\int_\pi^0 d\varphi \tfrac{r'^2}{r^2} = \\ &= \tfrac{Le^2}{\pi}\int_\pi^0 d\varphi \tfrac{\sin^2\varphi}{(1+e\cos\varphi)^2} \ ; \end{aligned} \tag{9.55}$$

integrating by parts, we get

$$I_r = -L + \frac{L}{\pi}\int_0^{\pi/2} d\varphi \left(\frac{1}{1+e\cos\varphi} + \frac{1}{1-e\cos\varphi}\right) \; ; \tag{9.56}$$

the integrals in this equation can be carried out by the substitution $\tan\varphi/2 = t$; finally, we get

$$I_r = -L + \frac{L}{\sqrt{1-e^2}} = -L + \sqrt{\frac{m\alpha^2}{2\mid E\mid}} \; . \tag{9.57}$$

It follows

$$1 - e^2 = \frac{I_\varphi^2}{(I_r + I_\varphi)^2} \tag{9.58}$$

and

$$E = -\frac{m\alpha^2}{2(I_r + I_\varphi)^2} \; . \tag{9.59}$$

The parameter p is given by $p = I_\varphi^2/m\alpha$. When m, α vary slowly, the eccentricity of the orbit remains unchanged, while its dimensions vary according to the parameter p.

According to the quantization rule given by equations (9.20) $I_\varphi = L$ is $\hbar$ times an integer $l+1$, $l = 0, 1, ...$, while I_r is $\hbar$ times an integer $n_r \geq 0$; it follows $I_r + I_\varphi = \hbar(n_r + l + 1)$ or $I_r + I_\varphi = \hbar n$, where $n = 1, 2, ...$ and $l = 0, 1, ...n-1$. The energy levels are $E_n = -m\alpha^2/2\hbar^2 n^2$, which coincide with the exact (quantum-mechanical) energy levels.[5]

9.5 Wavefunctions, central field

For the motion of a particle in a centrally-symmetric field the quantum-mechanical wavefunctions are $R(r)Y_{lm}(\theta, \varphi)$, where $R(r)$ are radial wavefunctions and Y_{lm} are the spherical harmonics, associated with the orbital motion along the spherical angles θ and φ; $Y_{lm} \sim P_l^{|m|}(\theta)\cdot$ $\cdot e^{im\varphi}$, where $P_l^{|m|}$ are the associated Legendre polynomials. For $\mid m \mid \simeq l \gg 1$ the associate Legendre polynomials are approximately

[5] A. Sommerfeld, *Atombau und Spektrallinien*, Vieweg&Son, Braunschweig (1939).

$P_l^l \simeq (\sin\theta)^l$; these functions are concentrated on the equatorial plane $\theta = \pi/2$; the angular momentum reduces practically to its z-component. In the opposite case $m = 0$ the associated Legendre polynomial are Legendre polynomials P_l; they are solutions of the equation

$$P_l^{''} + \cot\theta P_l^{'} + l(l+1)P_l = 0 \ , \tag{9.60}$$

which becomes

$$\chi^{''} + \left[(l+1/2)^2 + \frac{1}{4\sin^2\theta}\right]\chi = 0 \tag{9.61}$$

with the substitution $P_l = \chi/\sqrt{\sin\theta}$. The square bracket in equation (9.61) is $1/\lambda^2$, where λ is the wavelength; the quasi-classical condition $d\lambda/d\theta \ll 1$ gives $\theta l \gg 1$, $(\pi - \theta)l \gg 1$; the motion is quasi-classical for all θ's, except the vicinity of the poles; in this case equation (9.61) reduces to

$$\chi^{''} + (l+1/2)^2\chi = 0 \tag{9.62}$$

and $\chi = P_l\sqrt{\sin\theta} = A\sin[(l+1/2)\theta + \alpha]$, where A and α are two constants. On the other hand, for $\theta \ll 1$ equation (9.60) may be written as

$$P_l^{''} + \frac{1}{\theta}P_l^{'} + (l+1/2)^2 P_l = 0 \ , \tag{9.63}$$

whose solution is $P_l = J_0[(l+1/2)\theta]$, where J_0 is the Bessel function; for $l\theta \gg 1$ (and $\theta \ll 1$) we may use the asymptotic expression for the Bessel function

$$P_l \simeq \sqrt{\frac{2}{\pi l}}\frac{\sin[(l+1/2)\theta + \pi/4]}{\sqrt{\theta}} \ ; \tag{9.64}$$

hence $A = \sqrt{2/\pi l}$ and $\alpha = \pi/4$; the quasi-classical wavefunction is

$$P_l \simeq \sqrt{\frac{2}{\pi l}}\frac{\sin[(l+1/2)\theta + \pi/4]}{\sqrt{\sin\theta}} \tag{9.65}$$

and the corresponding (normalized) spherical harmonics are

$$Y_{l0} \simeq \frac{i^l}{\pi}\frac{\sin[(l+1/2)\theta + \pi/4]}{\sqrt{\sin\theta}} \ . \tag{9.66}$$

We turn now to the radial wavefunction. The radial wavefunction is a solution of the Schrodinger equation

$$\frac{1}{r^2}\frac{d}{dr}\left(r^2\frac{dR}{dr}\right)-\frac{l(l+1)}{r^2}R+\frac{2m}{\hbar^2}(E-V)R=0\ , \tag{9.67}$$

which becomes

$$\chi^{''}+\left[\frac{2m}{\hbar^2}(E-V)-\frac{l(l+1)}{r^2}\right]\chi=0 \tag{9.68}$$

with the substitution $R=\chi/r$; this is a one-dimensional Schrodinger equation with the effective potential

$$V_{eff}=V+\frac{\hbar^2 l(l+1)}{2mr^2} \tag{9.69}$$

(which includes the centrifugal potential associated with the angular momentum). The quasi-classical momentum is $p=\sqrt{2m(E-V_{eff})}$ and the conditions for quasi-classical motion must be analyzed for this momentum (or wavelength $\lambda=2\pi\hbar/p$), which includes the centrifugal potential (for $l\neq 0$). For attractive potentials the centrifugal potential dominates for small r and the momentum is $p\simeq l/r$; the quasi-classical motion is valid for $l\gg 1$, *i.e.* for peripheral states, where the particle moves far away from the origin. For the Coulomb potential $-\alpha/r$, $\alpha>0$, the critical region is of the order r, given by $\mid E\mid=\alpha/r$; it must be larger than the wavelength $\lambda\simeq\hbar/\sqrt{m\mid E\mid}$, *i.e.* $\mid E\mid\ll m\alpha^2/\hbar^2$, or $r\gg\hbar^2/m\alpha$ (Bohr radius).

9.6 Tunneling

A free-particle wavefunction of coordinate x is $\psi=e^{\frac{i}{\hbar}px}$, where p is momentum; we leave aside the amplitude; it can also be written as $\psi=e^{\frac{i}{\hbar}\int^x dx\cdot p}$, with an undetermined initial phase. This wavefunction can also be written as $\psi=e^{\frac{i}{\hbar}\int_a^x dx\cdot p}$ and $\psi=e^{\frac{i}{\hbar}\int_a^b dx\cdot p}e^{\frac{i}{\hbar}\int_b^x dx\cdot p}$, where $a<b$ are two distinct points; the momentum is $p=\sqrt{2mE}$, where m is the mass of the particle and E is the energy. Let us assume that a potential barrier $V(x)>E$ appears between a and b. For $x<a$ and

$x > b$ the wavefunctions preserve their form, but the wavefunction $\psi = Te^{\frac{i}{\hbar}\int_b^x dx\cdot p}$ for $x > b$ acquires an amplitude

$$T = e^{-\frac{1}{\hbar}\int_a^b dx|p|} \ , \tag{9.70}$$

where $\mid p \mid = \sqrt{2m[V(x) - E]}$. This is precisely what we expect from solving the quantum-mechanical problem for energy E with the potential barrier $V(x)$: for $x > b$ the wavefunction has the form $\psi = Te^{\frac{i}{\hbar}\int_b^x dx\cdot p}$ and for $x < a$ the wavefunction has the from $\psi = e^{\frac{i}{\hbar}\int_a^x dx\cdot p} + Re^{-\frac{i}{\hbar}\int_a^x dx\cdot p}$, where R is the reflection coefficient, T is the transmission coefficient and $\mid R \mid^2 + \mid T \mid^2 = 1$ for the conservation of the particle current; these amplitudes are determined from the continuity conditions at the borders of the barrier, with damped wavefunctions inside. The question arises of how accurate is equation (9.70) for T. It is obvious, from their construction, that the wavefunctions used for determining T are quasi-classical wavefunctions. Then, the regions of interest for $x < a$ and $x > b$ must be large in comparison with the wavelength $\lambda = 2\pi\hbar/\sqrt{2mE}$, *i.e.* λ must be small (high p and E) and the "wavelength" $\lambda = 2\pi\hbar/\sqrt{2m(V-E)}$ must vary slowly inside the barrier, *i.e.* $\frac{1}{\hbar}\int_a^b \sqrt{2m(V-E)}$ must be large: the potential barrier must be high in comparison with E and sufficiently wide, such that $\mid T \mid \ll 1$, as expected for small deviations from the classical behaviour; this is the "exponential accuracy" of the quasi-classical approximation. The effect described here is called the tunnel effect.[6] $\mid T \mid^2$ is the probability of penetration of the potential barrier; if we have n particles per unit length in the incident wavefunction ($x < a$), then we have $\mid T \mid^2 n$ particles per unit length in the transmitted wave; if we have a current density v in the incident wave, where v is the velocity, then we have a transmitted current density $\mid T \mid^2 v$.

We note that the tunneling is a stationary effect, for a fixed energy; it does not depend on the time and the transmission coefficient does not depend on the incident wavefunction. The tunneling may take place from a potential well to another (with energy conservation). It

[6] G. Gamow, "Zur Quantentheorie des Atomkernes", Z. Phys. **51** 204 (1928); "Zur Quantentheorie der Atomzertrummerung", Z. Phys. **52** 510 (1929); E. U. Condon and R. W. Gurney, "Quantum Mechanics and radioactive disintegration", Phys. Rev. **33** 127 (1929).

is often convenient to estimate the tunneling rate, *i.e.* the tunneling probability per unit time; it is$\mid T \mid^2 / \tau$, where $\tau \simeq \hbar/E$ is the attempt time of the particle to penetrate the potential barrier from the bound state.

There exists a generalization of the above formula for the tunneling coefficient. Indeed, let us assume that the "incident" wavefunction is ψ_1^0 for $x < a$; also, let us assume that for $x > a$ the wavefunction is $\psi_2^0 = A(x)e^{i\Phi(x)}$, where A is a real amplitude and Φ is a real phase. We can write this wavefunction as $\psi_2^0 = e^{i\phi(x)}$, where $\phi(x) = \Phi(x) - i\ln A(x)$, or $\psi_2^0 = e^{i\phi(a,x)}$, where $\phi(a,x) = \phi(x) - \phi(a)$; we assume a corresponding change in ψ_1^0, such that the continuity conditions be satisfied at $x = a$; similarly, we may introduce a corresponding pre-factor in ψ_2^0. We can also write $\psi_2^0 = e^{i\phi(a,b)}e^{i\phi(b,x)}$. Now, let us introduce a potential barrier between a and b; the wavefunction ψ_1^0 acquires an additional reflected wavefunction, the wavefunction between a and b changes, but we are not interested in it; we are interested in the wavefunction for $x > b$ (outside the barrier), which is $\psi_2 = Te^{i\phi(b,x)} = T\psi_2^0$, where $T = e^{i\phi(a,b)}$; this factor is changed by the potential barrier, such that we may write

$$\phi(a,b) = \int_a^b dx \frac{\partial \phi}{\partial x} = \int_a^b dx \left(\frac{\partial \Phi}{\partial x} - i \frac{\partial \ln A}{\partial x} \right) \; ; \tag{9.71}$$

the coefficient T computed this way is the quantum-mechanical transmission coefficient. The transition amplitude from ψ_2^0 to ψ_2 is T. The computation of T requires in fact the solution of the quantum-mechanical problem inside the barrier. It may happen that $\partial\phi/\partial x$ has a zero in the complex plane; then, the estimation of the integral in equation (9.71) can be made by means of a contour passing through this point, which amounts to the saddle-point method. If A varies slowly in the barrier and Φ is $S/\hbar$, the above formula reduces to the quasi-classical tunneling, with

$$\phi(a,b) = \Phi(a,b) = \int_a^b dx \cdot p \; ; \tag{9.72}$$

the transmission coefficient computed this way may be termed the quasi-classical transmission coefficient.

Similar considerations can be made for wavefunctions which depend on the time too ($\psi(t,x)$); there appears a difficulty in this case, related to the integration in equation (9.71) from (t_0, a) to (t, b); in general, such an integral depends on the path of integration.[7] However, if the transition process does not depend practically on the position, *i.e.* if it depends only on the time, we can perform, in principle, the integral

$$\phi(t_0,t) = \int_{t_0}^{t} dt \left(\frac{\partial \Phi}{\partial t} - i \frac{\partial \ln A}{\partial t} \right) \quad , \tag{9.73}$$

which gives the quantum-mechanical transmission coefficient; in the quasi-classical approximation it reduces to

$$\phi(t_0,t) = \Phi(a,b) = - \int_{t_0}^{t} dt \cdot E \ . \tag{9.74}$$

We note that the quasi-classical behaviour is necessary inside the barrier, according to these formulae, but, in this case, it implies in fact a quasi-classical behaviour everywhere.

The tunneling process as described above is a stationary process. However, there is a transient regime when the particle is propagating, reflected, transmitted and propagating again; in this case there exists a delay time, which is the time spent by particle inside the barrier ("dwelling" time). It can be estimated from the phase variation $(\partial\phi/dx)dx + (\partial\phi/\partial t)dt = 0$, whence we get the local phase velocity $v = \mid (\partial\phi/\partial t)/(\partial\phi/dx) \mid$ inside the barrier; the delay time is an integral which implies $d\tau = dx/v$, but we have again difficulties in estimating such an integral which depends on the integration path. In the quasi-classical case the phase velocity is $v = \mid E/\sqrt{2m(V-E)} \mid$ and the delay time is $\tau = \int_a^b dx \mid p \mid /E = \hbar \mid \ln T \mid /E$; it is much longer than the time $d/\sqrt{2E/m}$ needed by the free particle to pass from a to b through a barrier of width d, since $E \ll V/2$.

Equation (9.70) has been applied originally to the nuclear alpha disintegration, where the spontaneous decay of the alpha particle proceeds by penetrating the nuclear Coulomb potential barrier. For a Coulomb

[7] This way we may arrive at Feynman path integrals (R. P. Feynman, "Space-time approach to non-relativistic quantum-mechanics", Revs. Mod. Phys. **20** 367 (1948)).

potential $V = \alpha/r$ in equation (9.70) we get the transmission coefficient

$$T = e^{-\frac{\alpha}{\hbar}\sqrt{2m/E}\left[\arccos\sqrt{aE/\alpha}-\sqrt{(aE/\alpha)(1-aE/\alpha)}\right]} ; \qquad (9.75)$$

for $a = 0$, the transmission coefficient is $T = e^{-\pi\alpha/\hbar v}$, where v is the velocity of the particle. The same equation is used for atom ionization in a uniform electric field $\mathcal{E}$. Indeed, the potential barrier is 0 for $x < 0$ and $V_0 - q\mathcal{E}x$ for $x > 0$ in that case, where V_0 is the height of the potential barrier and q is the electron charge; the penetration goes from $x = 0$ to $x_0 = (V_0 - E)/q\mathcal{E}$, where E ($< V_0$) is the energy of the particle; we get

$$T = \frac{2\sqrt{2m}}{3\hbar q\mathcal{E}}(V_0 - E)^{3/2} , \qquad (9.76)$$

a result valid for low fields $\mathcal{E}$ ($T \ll 1$); for a potential well, we should subtract V_0 from $V_0 - E$ and the remaining $-E$ is the binding energy E_b of the electron; equation (9.76) can also be written as

$$T = \frac{2E_b^{3/2}}{3(q\mathcal{E}a)(\hbar^2/2ma^2)^{1/2}} , \qquad (9.77)$$

where a is the radius of the bound state; the validity condition requires $T \ll 1$. Originally, equation (9.77) has been applied to the hydrogen atom,[8] but it may be extended to heavy atoms (and similar bound states).[9]

If a particle moves in two potential wells separated by a potential barrier V, then the energy level E corresponding to each well is split (it is degenerate). The splitting can be estimated from $E = (\psi, H\psi)$ by $\delta E \simeq ET$, because $\delta\psi$ is of the order $T\psi$.

---

[8]C. Lanczos, "Zur Theorie des Starkeffektes in hohen Feldern", Z. Phys. **62** 518 (1930); "Zur Intensitatsschwachung der Spektrallinien in hohen Feldern", Z. Phys. **68** 204 (1931); J. R. Oppenheimer, "Three notes on the quantum theory of aperiodic effects", Phys. Rev. **31** 66 (1928); H. A. Bethe and E. E. Salpeter, *Quantum Mechanics of One- and Two-Electron Atoms*, Springer, Berlin (1957).

[9]Y. N. Demkov and G. F. Drukarev, "Decay and polarizability of negative ions in electric fields", Sov. Phys.-JETP **20** 614 (1965) (ZhETF **47** 918 (1964)).

9.7 Chemical reactivity

Let us consider a quantum-mechanical bound-state of a particle placed at the origin. An external perturbation (like an electric field) creates a potential barrier on one side of the bound state, along a coordinate. In its motion along this direction the particle changes its quantum-mechanical phase, with a local phase velocity. When the front of this wave arrives at the border of the potential barrier, the particle is reflected back into the bound state with amplitude R and penetrates forth through the barrier with amplitude T; then it is reflected back by the other side of the bound state, with the same amplitude, and another fraction T of the particle tunnels again through the barrier. Let t_0 be the time needed for the particle phase to oscillate back and forth inside the bound state; then, a long time t may be written as

$$t = \tau + 2t_0 + 2t_0 + ... = \tau + 2nt_0 \ , \tag{9.78}$$

with an uncertainty in the initial time τ of the order $2t_0$. For a quantum-mechanical tunneling the delay time in the barrier is small, and we may neglect it. The probability of finding the particle outside the barrier is

$$\begin{gathered} P = |T|^2 + |R|^2 |T|^2 + |R|^4 |T|^2 + ... = \\ = |T|^2 \tfrac{1-|R|^{2(n+1)}}{1-|R|^2} = 1 - |R|^{2(n+1)} = 1 - |R|^{t/t_0+(2-\tau/t_0)} \ ; \end{gathered} \tag{9.79}$$

averaging over τ for small T leads to

$$P \simeq 1 - |R|^{t/t_0} = 1 - e^{\ln|R|t/t_0} \ , \tag{9.80}$$

or $P = 1 - e^{-Ct}$, where $C = -\ln \mid R \mid /t_0$. This is the well-known law of chemical reactivity and of e^{-Ct}-decay for the survival probability $1 - P(t)$; C is the rate constant of the chemical reaction.[10]

The wavefunction of the n-th transmitted wave is $\mid T \mid\mid R \mid^n \times$ $\times e^{-\frac{i}{\hbar}E(t-2nt_0)+\frac{i}{\hbar}px}$, where E is the energy and p is the momentum of the particle, or $\mid T \mid\mid R \mid^n e^{-\frac{i}{\hbar}Et+\frac{i}{\hbar}px}$, since $E = (2\pi\hbar/2t_0)n$. The

[10] M. Apostol, "On the rate of the chemical reactions and the teleportation of the wavepackets", J. Theor. Phys. **74** (2001).

emission is a statistical process, which means that each wavefunction carries a random phase φ_n. It follows that the probability of emission is given by

$$| T |^2 \sum_{n,m} | R |^{n+m} e^{i(\varphi_n - \varphi_m)} = | T |^2 \sum_n | R |^{2n} \ , \tag{9.81}$$

which is indeed the probability given by equation (9.79). Equation (9.81) shows that the emission process is incoherent. The quantum-mechanical tunneling described here is practically instantaneous (as if the waves are teleported through the barrier) and the phases φ_n account for the uncertainty in localization of the particle.

For a complex bound-state, or for a strong external perturbation, the particle may be a localized wavepacket, and the tunneling may be a quasi-classical tunneling. In this case the delay time spent by the particle inside the barrier should be included in t_0 and the localization probability of the emitted particle is

$$| T |^2 \sum_n | R |^{2n} \delta[x - v(t - 2nt_0)] \ , \tag{9.82}$$

where x is the coordinate of the motion and v is the group velocity; the total probability of finding the particle outside the barrier is obtained by integrating over x; we get again the probability given by equation (9.79).

9.8 Particles and wavepackets

The experimental realization (preparation) of the state of the particles is of utmost importance for understanding the meaning of the Quantum Mechanics. For brevity, we talk about electrons here, but the discussion applies to other particles too.

The quantum-mechanical behaviour is more general than the quasi-classical or the classical motion, so we start with it; it implies small amounts of mechanical action and relates to particles at the atomic level. The Planck constant $\hbar$ is the lower bound for the mechanical action; there is not a well-defined upper bound for the quantum-mechanical action, which would indicate a clear-cut transition to the classical behaviour; the transition to the classical behaviour is gradual. The classical behaviour is a limiting case of the quantum-mechanical behaviour. Atomic particles have a quantum-mechanical behaviour not only over small spatial extensions, but also in large, macroscopic bodies, like superfluid or superconducting bodies, whose motion implies small amounts of mechanical action.

Let us first consider an electron in a bound state. In general, there exist many electrons in a bound state, and their wavefunction is entangled, in the sense that it cannot be decomposed in superpositions of products of single-electron wavefunctions; the many-particle wavefunctions of identical particles are either symmetric or antisymmetric, according to the spin of the particles (bosons for integral spin, fermions for half-integral spin). This entanglement imposes a serious limitation upon the many-particle wavefunctions. It means that we do not know in fact the many-particle wavefunction, and we only may do some approximations. This limitation might not be absolute, in principle; it may only relate to our incapacity of solving the Schrodinger equation for many particles; however, in practice, this limitation has an absolute character. As concerns the dynamics of the assembly we may not be very much interested in its ground-state, but, instead, in its excited states, or in states which imply small deviations from the ground-state; these states, also, are not easily available for a few-electron (few-body) assemblies. If the number of electrons is very large (many-body problem), the low-lying states in the vicinity of the ground-state are available (elementary excitations), either as quasi-particles (introduced by Landau for fermions) or collective excitations; they imply small deformations of the ground-state. For the elementary excitations we have (approximate) one-particle wavefunctions (lifetime included). The mean-field concept introduced by Bohr provides a reasonably satisfactory picture of single-particle states; highly excited one-particle states are described satisfactorily by the mean-field concept, even for a few-body assemblies. We assume here a mean-field model for the one-electron states in a bound state.

A quantum-mechanical one-electron state is described by a wavefunction of the form $\psi(t,\mathbf{r}) = A(t,\mathbf{r})e^{i\Phi(t,\mathbf{r})}$, with a real amplitude A and a real phase Φ; it can be written as $e^{i\phi}$, where $\phi = \Phi - i\ln R$.

The quantum-mechanical wavefunction looks like a field. Since the physical quantities are described by bilinear forms of the wavefunctions, we should admit an imaginary phase besides the amplitude of the wavefunction. This amounts in fact to having two fields in the wavefunction. A local and instantaneous quasi-energy may be defined by $E(t, \mathbf{r}) = -\hbar\partial\phi/\partial t$ and a local and instantaneous momentum $\mathbf{p}(t, \mathbf{r}) = \hbar grad\phi$; hence, we may have a period $T(t, \mathbf{r})$ of the motion and a wavelength $\lambda(t, \mathbf{r})$, both depending on the time and position. These quantities are, in general, complex, and for comparing them with real quantities we should use both their real and imaginary part. An instantaneous and local quantum-mechanical state can be defined, with uncertainties T in time and λ in space. For shorter times and wavelengths the state does not exist and the Quantum Mechanics is blocked. Indeed, since $\Delta p \Delta x \gg \hbar$ and $p = \hbar/\lambda$ (where the momentum p is associated to the coordinate x), we have $\Delta p = \hbar\Delta\lambda/\lambda^2$ and $\Delta\lambda\Delta x \gg \lambda^2$, *i.e.* $\Delta x \gg \lambda$; and similarly for the period. In its excursion over the finite dimension of the bound state the phase may experience many periods and many wavelengths; if the excursion time τ is of the order nT and the dimension of the bound state d is of the order $n\lambda$, where n is an integer (it should be an integer, for the definiteness of the states), then, according to their definition given above, the energy is of the order $\hbar/T \simeq n(\hbar/\tau)$ and the momentum is of the order $\hbar/\lambda \simeq n(\hbar/d)$; we can see that for large τ and d the energy levels and the quantum-mechanical states are dense; of course, there may exist many T and λ, and motion is possible also for multiples of the fundamental T and λ, which means that n in the above formulae is independent of τ/T or d/λ. The particles exist (more rigorously, their quantum-mechanical states exist) over characteristic wavelengths at any point in space inside their bound states and during their characteristic periods at any time. We may define an instantaneous and local phase velocity of the wavefront $\phi = const$, by $v_{ph} = |(\partial\phi/\partial t)/(| grad\phi ||$, along the direction of the $grad\phi$; also, we may define a group velocity by $\mathbf{v}_g = \partial E/\partial \mathbf{p}$, perpendicular to the curves $E = const$ (distinct from the direction and the magnitude of the phase velocity), with E and $\mathbf{p}$ defined above. Likewise, we may define the excursion time as $\int dl/v_{ph}$, along the path perpendicular to the wavefront. These paths may be viewed as quantum-mechanical "trajectories", and the particle is simultaneously on all of them (this is a quantum-mechanical ubiquity). Using the above definitions we have $i\hbar\partial\psi/\partial t = E\psi$ and $-i\hbar grad\psi = \mathbf{p}\psi$, or the association $E \leftrightarrow i\hbar\partial/\partial t$ and $\mathbf{p} \leftrightarrow -i\hbar grad$; it follows that we should have $i\hbar\partial\psi/\partial t = -(\hbar^2/2m)\Delta\psi + V\psi$, which is the Schrodinger equation for a particle with mass m in the potential V. An important simplification is obtained for a wavefunction of the form $\psi(t,\mathbf{r}) = e^{-\frac{i}{\hbar}Et}\varphi(\mathbf{r})$, where $\varphi(\mathbf{r}) = A(\mathbf{r})e^{i\chi(\mathbf{r})}$, which is the stationary case, with well-defined energies, given by the eigenvalue Schrodinger equation $-(\hbar^2/2m)\Delta\varphi + V\varphi = E\varphi$, with V independent of time (conservative potential). The period $T = 2\pi\hbar/E$ is then real and constant, the excursion time τ is nT; the wavelength remains local and complex.

A special situation occurs in a crystalline solid, where there exists a lattice constant, say, a, along a direction, and the electron-ion potential is periodic with periodicity a along that direction. In that case we note that there exists a fundamental wavelength a, so the momentum should be of the form $p_n = n(\hbar/Na)$, where N is the number of ions along that direction, and n is limited by $\pm N/2$; in addition, when n is close to $\pm N/2$ the ionic potential becomes effective and the degenerate states moving in opposite directions get mixed, their energy is split and there appears an energy gap; that means that there may exist an amplitude $u_m(x)$ for the wavefunction $e^{\frac{i}{\hbar}p_n x}$, with the periodicity a and a state label m generated by the ionic potential. This is the so-called Bloch theorem for crystalline solids and the picture of the energy bands in crystalline solids.[11] It is worth noting here that the high-energy electron states are defined with uncertainty a at any point along the solid; the high-energy electrons are defined within the distance a at any point of the solid. It may happen that the number of electrons in the upper band is smaller than the number of ions, then they are defined over a distance a of the order of their mean separation distance. In amorphous solids, although the wavefunctions are different, the mean inter-ionic distance or the mean inter-electron distance defines the electrons.

If the characteristic wavelength λ is sufficiently small in comparison with the distance d available for the motion of the particle, which implies a sufficiently high energy and momentum, then the particle

[11] F. Bloch, "Uber die Quantenmechanik der Elektronen in Kristallgittern", Z. Phys. **52** 555 (1928).

may be found with practically the same probability everywhere along the distance d; this implies that the wavefunction is $e^{i\Phi}$, *i.e.* the amplitude is a constant; Schrodinger's equation for such a wavefunction reduces to the Hamilton-Jacobi classical equation for the mechanical action $S = \hbar\Phi$. This is the quasi-classical behaviour. The motion is practically classical, except for the existence of a constant, undetermined phase φ in the wavefunction $e^{\frac{i}{\hbar}S+i\varphi}$, which accounts for the undetermined place of the particle along the distance d. The existence of a small wavelength λ implies the existence of a small period T, with a slow variation in time at most. The phase φ may also account for the initial time; it defines the incoherent regime of the quasi-classical motion. We can see that the border between quantum-mechanical behaviour and the quasi-classical behaviour is characterized by the existence of a non-trivial (variable) amplitude of the wavefunction.

The short characteristic wavelength and period of a quasi-classical motion define, in principle, a quantum-mechanical state. However, if the characteristic wavelength and period are much shorter than the distance and the duration over which we consider the motion, *i.e.* if the particle is quasi-classical, it is very likely to exist also other characteristic wavelengths and periods which differ slightly from the former and from each other; it is very likely to exist dense states for the quasi-classical particles, like, for instance, in macroscopic bodies. Then, for any state we should consider a superposition of the dense states lying in the vicinity of that state, with the same phases φ, since the states belong to the same particle; this superposition is a wavepacket. Indeed, let us write for the mechanical action

$$S = S_0 - vpt + px \ , \tag{9.83}$$

where $S_0 = -E_0 + p_0 x$ and $v = \partial E/\partial p\mid_0$ is the group velocity; p denotes the deviation of the momentum from p_0; we limit ourselves to the linear expression in equation (9.83) with the accuracy $\Delta p \simeq p_0$. The states are labelled by p and the superposition is

$$\begin{gathered}\psi = Ce^{\frac{i}{\hbar}S_0}\int dpe^{\frac{i}{\hbar}(x-vt)p} = Ce^{\frac{i}{\hbar}S_0}\frac{\sin\frac{(x-vt)\Delta p}{2\hbar}}{\frac{(x-vt)}{2\hbar}} \rightarrow \\ \rightarrow Ce^{\frac{i}{\hbar}S_0}(2\pi\hbar)\delta(x-vt) \ ;\end{gathered} \tag{9.84}$$

we can see that the wavefunction is localized over a distance $\Delta x \simeq \hbar/\Delta p \simeq \lambda_0$, where λ_0 is the wavelength corresponding to p_0; the wavefunction given by equation (9.84) tends to a delta function in the limit $\lambda_0 \simeq 0$; the normalization of the wavefunction ψ implies

$$\psi = \frac{e^{\frac{i}{\hbar}S_0}}{\sqrt{\lambda_0}}\delta(x - vt) \ ; \tag{9.85}$$

the particle is localized over distance λ_0 (and time $T_0 = \lambda_0/v \simeq \hbar/\Delta E$), without knowing its position (or time), in accordance with the phase φ. The quasi-classical particles may be represented by such wavepackets, which describe, practically, classical particles; except for the phase factor $e^{\frac{i}{\hbar}S_0}$, which is reminiscent of the quantum-mechanical behaviour of the particle. It plays the main role in tunneling, with a small quantum-mechanical contribution, as expected. It is worth noting that the linear expression used above for deriving equation (9.84) is not a limitation, as long as we use its correct uncertainty. Of course, the above considerations apply to the motion in the whole space, as long as the motion is separable.

Let us suppose that we wish to create a beam of electrons. This is done by heating locally a solid, thus creating a local plasma, and applying an electric potential to extract the electrons through an aperture. Usually, the aperture is much wider than the mean separation distance which is characteristic for the electrons in plasma, or in solids; we may view it as being of the order of atomic distances. The extraction is efficient for the most energetic electrons, which, in these conditions are quasi-classical, experiencing a dense multitude of states; along transverse directions (with respect to the extracting electric field), they are localized over distances of the order of the mean separation distance; this plays the role of the characteristic wavelength λ_0; of course, they are extracted at various transverse positions on the opening of the aperture. The extracted electrons follow an almost classical trajectory, with an undetermined position along this trajectory, reminiscent of their quantum-mechanical nature; we may use for them wavefunctions of the form given by equation (9.85); the electron beam may be viewed as consisting of a set of parallel electronic threads, of an incoherent nature. For each thread we may use a wavefunction of the form $e^{\frac{i}{\hbar}S_0}$; the extension of this wavefunction to the whole beam may induce the misleading perception that all the initial phases of the electrons are the same, *i.e.* they are in a coherent regime, which usually, is not the case. The coherent regime may establish itself for a collection of particles whose collective interaction implies very low amounts of mechanical action; the small interaction is able to correlate the motion into a coherent phase.

9.9 "Philosophical" problems

The basic trouble of the Quantum Mechanics is the particle-wave duality.

Heisenberg's matricial Quantum Mechanics associates a mechanical state to any body, determines the measurable physical quantities as matrix elements of the associated operators, describes their time evolution and realizes that the quantum-mechanical particles have not a trajectory; the states have probabilities and the quantum jumps are described by transition probabilities; the matricial mechanics works with mechanical quantities like position, momentum, energy and conveys the perception that the quantum-mechanical bodies are particles, even without a well-defined position.

Schrodinger's Quantum Mechanics associates to the quantum-mechanical bodies a wavefunction, subject to a wave equation, realizes that the quantum-mechanical states are associated with solutions of eigenvalue equations and determines matrix elements as scalar products of wavefunctions; the wavefunction is a wave of probability; the wave mechanics conveys the perception that the quantum-mechanical bodies are waves. The two Quantum Mechanics are two different languages with a dictionary to pass from one another. It may appear the question of what the quantum-mechanical bodies are, ultimately: particles or waves? The situation looks as if it would be overdetermined, with two contradictory solutions.

This question may be resolved by noticing that in science we never talk about what the world is, but instead of how the world behaves, how it moves. Although in Classical Mechanics we say the body is a particle or a wave, we always mean that body has momentum, energy, position, etc. Science is not the world, it is about the world, specifically how we talk about the world. More interesting are the new limitations the Quantum Mechanics brings, like the uncertainty in determining position, momentum, energy, time, and the new, more abstract concepts it introduces, like the quantum-mechanical state as a collection of measurable quantities and the wavefunction as a wave of probability, statistical behaviour of the quantum-mechanical bodies. "It is wrong to think that the task of Physics is to find out how Nature is. Physics concerns what we say about Nature" (Bohr).

An extended wavefunction of an electron falling on a photographic screen passes into a localized wavefunction of that electron immersed in the screen. This collapse of the wavefunction is viewed as another trouble of the Quantum Mechanics. However, if we view the wavefunction not as a material wavefunction but as a probability wave, the collapse does not seem to appear so troubling anymore. It becomes simply a transition from one state to another, governed by an interaction.

In the double-slit experiment the particle passes simultaneously through both slits, as if it would be a wave; and it gives interference on the screen. However, if we measure the position of the particle, we find it near one, or another, slit. This circumstance is again viewed as troubling. We may expect a wave to pass through the both slits simultaneously, while, when measuring the position, the extended wave collapses in one place, or another, never in two places at the same time, because the final waves are localized in the vicinity of each place and they are not superposed. More interesting is how a probability wave behaves like a physical, material wave, and diffracts and interferes; this is so because the probability wave obeys a wave equation (Schrodinger's equation) which is constructed with physical quantities (position, time, momentum, energy, mass) specific to the material waves. This again shows the more abstract nature of the concept of waves of probability.

The state of many identical particles is defined by the sum of the physical quantities of each particle; the wavefunction of such an assembly is entangled. We may measure a physical quantity in such an extended entangled wavefunction of, say, two particles, at one place; we find a state of one particle (the probability of finding the two particles in that state is very small). At that moment, the state of the other particle is determined, according to the summation rule of their physical quantities. This circumstance is viewed as the transmission of information at distance instantaneously, which would be an unphysical situation; it would be a teleportation, a spooky action at a distance, or a quantum telepathy; it is called the Einstein-Podolsky-Rosen paradox.[12] This is true. However, it is not so troubling, since we do not transmit the information through a physical means; the information is distributed over the whole space occupied by the entangled wavefunction, according to its preparation; when we measure the wavefunction at one place, we measure it everywhere.

A cat imprisoned in a black box can be killed at any random moment by a mechanism triggered by, say, a random nuclear disintegration. The wavefunction of the cat is a superposition of two states, dead and alive. However, when we open the box, we find the cat in one state, either alive, or dead. Wigner put his friend inside the box, next to the cat; the friend sees the cat dead or alive, not a superposition of dead and alive. This is known as Schrodinger's paradox of the quantum cat.[13] It is true. The physical quantities make sense only when measuring them; the measuring means give well-defined results; the superposition of states is not a measurement.

The difficulties in understanding the Quantum Mechanics gave rise to various "interpretations" of this scientific discipline. The uncertainty relations, the complete set of (compatible) measurable quantities, the particle-wave duality of the language, the statistical nature of the quantum-mechanical behaviour are known as the complementary picture or the Copenhagen interpretation; it expresses some views, often contradictory, mainly of Bohr and Heisenberg. Although Bohr clarified many aspects of the Quantum Mechanics, he also contributed much to obscuring other aspects (or the same), in accordance with his inappropriate belief that language would not be adequate, not suitable (and probably not desirable) for science. Similarly, Heisenberg's too much talking helped to establish the notion that there would be a need for some "interpretation" of the quantum-mechanical concepts. There are many so-called interpretations of the Quantum Mechanics,

[12] A. Einstein, B. Podolsky and N. Rosen, "Can quantum-mechanical description of physical reality be considered complete?", Phys. Rev. **47** 777 (1935).

[13] E. Schrodinger, "Die gegenwartige Situation in der Quantenmechanik", Naturwissenschaften **23** 807 (1935).

one more obscure than other. For instance, the assembly interpretation assumes that the statistical behaviour of the quantum-mechanical bodies relates only to statistical ensembles, not to individual bodies;[14] the guiding-wave theory assumes that the particles are guided by some pilot waves, through hidden variables;[15] the many-worlds interpretation claims obscurely that the measurements take us into one or other universe, these universes being parts of a multiverse;[16] the transitions would appear through an obscure mechanism of decoherence;[17] etc, etc. As long as we understand the Quantum Mechanics there is no need for an "interpretation" of that understanding; the "interpretation" is precisely our understanding. However, the difficulties in understanding the Quantum Mechanics should not be underestimated ("if Quantum Mechanics hasn't profoundly shocked you, you haven't understood it yet", Bohr; "I think I can safely say that nobody understands Quantum Mechanics", Feynman); they are sometimes as substantial as to give rise to perfectly trivial products as the so-called Bell's theorem,[18] which states, of course rigorously, that the Quantum Mechanics is distinct from the Classical Mechanics.

9.10 Reflection above the barrier

The tunneling described until now in this chapter is the quantum-mechanical tunneling through a classically inaccesible region, computed quasi-classically. Equations (9.71) and (9.73) provide a way to estimate the general quantum-mechanical tunneling; by means of equations (9.72) and (9.74) we can estimate this tunneling quasi-classically. We may view such a tunneling as a transition. The transitions are quantum jumps; as such, they are described by quantum-mechanical transition probabilities. In some cases, within the quasi-

[14] M. Born, "Zur Quantenmechanik der Stossvorgange", Z. Phys. **37** 803 (1926).

[15] L. de Broglie, Solvay Conference, 1928; D. Bohm, "A suggested interpretation of the quantum theory in terms of hidden variables, I and II" Phys. Rev. **85** 166 (1952).

[16] H. Everett, "Relative state formulation of quantum mechanics", Revs. Mod. Phys. **29** 454 (1957).

[17] H. D. Zeh, "On the interpretation of measurement in quantum theory", Found. Phys. **1** 69 (1970).

[18] J. Bell, "On the Einstein-Podolsky-Rosen paradox", Physics **1** 195 (1964).

classical approximation, the quantum jumps may look as classical motion through a complex path. Of course, this is a quantum-mechanical description.

Indeed, in general, for a stationary transition, according to equation (9.72), we solve the classical equations of motion for the motion from q_1 to q_2, and get the motion path; the path is complex, in general, and may pass through a complex q_0, given by the conservation laws, where the transition from one state to another occurs formally;[19] then the transmission coefficient from q_1 to q_2 is given by

$$T \simeq e^{-\frac{1}{\hbar}Im[S_1(q_1,q_0)+S_2(q_0,q_2)]} \; ; \tag{9.86}$$

if there are several points q_0, we choose the point that makes the largest T (among the q_0-points, the singularities of the potential must be considered too); if q_0 is real, but lies in the classically inaccessible region of motion, equation (9.86) becomes equation (9.70). Equation (9.86) can also be applied for reflection coefficients R. It is worth noting that the form S_2 of the mechanical action in the region beyond the transition point may be different from the form S_1 behind the transition point, which represents a more general circumstance.

Let us estimate the reflection coefficient above a potential barrier; *e.g.*, let $V(x) = -kx^2$ be a potential barrier, with a constant k, and let a particle with mass m move with energy $E > 0$ (above the barrier); we can see that $E = V(x)$ for $x_0 = i\sqrt{E/k}$. The reflection coefficient is

$$R = e^{-\frac{2}{\hbar}Im\int_x^{x_0} pdx} = e^{-\frac{2}{\hbar}Im\int_x^{x_0}\sqrt{2m(E+kx^2)}dx} = e^{-\frac{\pi E}{2\hbar}\sqrt{2m/k}} \tag{9.87}$$

(and the reflection probability is R^2); note that x_0 is a branch point for p above, which suggests an equivalent integration in the complex plane. The momentum is very large for large values of x. The lower limit x in equation (9.87) is immaterial (the phase is real there). We note that equation (9.86) implies a passing through a "potential barrier", either as a reflection or a transmission, and the quasi-classical approximation should be satisfied for this transition.

[19] L. Landau, "Zur Theorie der Energieubertragung bei Stossen", Phys. Z. Sowjet. **1** 88 (1932); L. Landau and E. Lifshitz, *Course of Theoretical Physics*, vol. 3, *Quantum Mechanics*, Elsevier (1977).

The reflection (and trasmission) above a potential barrier has an important general feature. Let us consider a potential barrier with distinct limits at $\pm\infty$ and the asymptotic form of the (plane) waves (quasi-classical waves)

$$\begin{aligned} &A_1 e^{ik_1 x} + B_1 e^{-ik_1 x} \quad , \quad x \to -\infty \\ &A_2 e^{ik_2 x} + B_2 e^{-ik_2 x} \quad , \quad x \to \infty \quad , \end{aligned} \tag{9.88}$$

where $k_{1,2,}$ are the corresponding wavevectors; since these waves belong to the same solution of the Schrodinger equation, A_2 and B_2 should be a linear combination of A_1 and B_1; let $A_2 = \alpha A_1 + \beta B_1$. The same solution can be written in complex conjugate form,

$$\begin{aligned} &A_1^* e^{-ik_1 x} + B_1^* e^{ik_1 x} \quad , \quad x \to -\infty \\ &A_2^* e^{-ik_2 x} + B_2^* e^{ik_2 x} \quad , \quad x \to \infty \quad , \end{aligned} \tag{9.89}$$

where we should have $B_2^* = \alpha B_1^* + \beta A_1^*$, or $B_2 = \alpha^* B_1 + \beta^* A_1$. The current conservation implies

$$k_1 \left(\mid A_1 \mid^2 - \mid B_1 \mid^2 \right) = k_2 \left(\mid A_2 \mid^2 - \mid B_2 \mid^2 \right) \quad ; \tag{9.90}$$

making use of the expressions of A_2 and B_2 found above, we get

$$\mid \alpha \mid^2 - \mid \beta \mid^2 = k_1/k_2 \ . \tag{9.91}$$

For one process of reflection and transmission, say, from $-\infty$ to ∞, we put $B_2 = 0$, *i.e.* $B_1/A_1 = -\beta^*/\alpha^*$, whence the reflection coefficient is given by

$$R^2 = \mid B_1/A_1 \mid^2 = \mid \beta/\alpha \mid^2 \ ; \tag{9.92}$$

from equation (9.91) we have $1 - R^2 = (k_1/k_2)/ \mid \alpha \mid^2$, which is the transmission coefficient; $B_1/A_1 = -\beta^*/\alpha^*$ is the reflection amplitude. For the process going from ∞ to $-\infty$ we set $A_1 = 0$ and the reflection amplitude is $A_2/B_2 = \beta/\alpha^*$; the two reflection amplitudes may differ by a phase.

According to equation (9.86) the reflection amplitude is given by

$$-\beta^*/\alpha^* = -ie^{\frac{2i}{\hbar}\left(\int_x^{x_0} pdx + px\right)} \quad , \quad x \to -\infty \ . \tag{9.93}$$

The reflection above a barrier can be computed also by means of the perturbation theory, viewing the reflection as a transition. The transition probability is given by the general formula

$$dw = \frac{2\pi}{\hbar} |V_{fi}|^2 \delta(E_f - E_i) d\nu_f \quad , \tag{9.94}$$

with usual notations; the transition is in fact a scattering, and the perturbation theory gives the Born cross-section. It is important to note that within the perturbation theory the initial and final wavefunctions are unperturbed, such that we consider the initial wavefunction $e^{\frac{i}{\hbar}px}/\sqrt{v}$ and the final wavefunction $e^{\frac{i}{\hbar}p'x}/\sqrt{v}$, where $p' = -p$ and $v = p/m$ is the velocity; these wavefunctions are normalized to the unit current density; the important point is that these wavefunctions extend over the whole space (as unperturbed wavefunctions). The density of final states is $d\nu_f = \mid p' \mid /2\pi\hbar$, and we integrate over p'; we get the reflection probability

$$R^2 = w = \frac{m^2}{\hbar^2 p^2} \left| \int_{-\infty}^{+\infty} dx V(x) e^{\frac{i}{\hbar}2px} \right|^2 \quad . \tag{9.95}$$

In one dimension the validity condition $\mid V \mid \ll \hbar^2/ma^2$ for the perturbation theory for a short-range potential, with a width a, is not valid anymore, since the wavefunction given by the Born approximation becomes infinite (a similar situation occurs in two dimensions); we should use the solution of the Schrodinger equation for motion along a coordinate; it follows $\mid V \mid \ll (\hbar^2/ma^2)ka$, or $\mid V \mid \ll \hbar v/a$, where $k = p/\hbar$; in addition $ka \leq 1$ to ensure that the integral in equation (9.95) is not exponential. If we compare this equation with equations (9.86) and (9.87), we can see that there is an important difference between the two results.

If there is a discontinuity in the slope of the potential barrier, of the form $V = -F_1 x$ for $x < 0$ and $V = -F_2 x$ for $x > 0$, the main contribution to the reflection above the barrier may come from this discontinuity; equation (9.95) gives $R^2 = (\hbar^2 m^2/16p_0^6)(F_1 - F_2)^2$, where $p_0 = p(x = 0)$.[20]

[20] L. Landau and E. Lifshitz, *Course of Theoretical Physics*, vol. 3, *Quantum Mechanics (Non-relativistic Theory)*, Elsevier, Oxford (2005).

An interesting application of the transition probability given by equation (9.86) is the disintegration of the deuteron in the Coulomb field of a heavy nucleus.[21] The main contribution to the disintegration rate is brought by the zero orbital momentum, such that the motion is one-dimensional. At some point r_0, the conservation laws of momentum and energy read

$$\begin{aligned} mv_p + mv_n &= 2mv_d \ , \\ \tfrac{1}{2}mv_p^2 + \tfrac{1}{2}mv_n^2 &= mv_d^2 - \varepsilon \ , \end{aligned} \tag{9.96}$$

where $v_{p,n,d}$ are the velocities of the proton, the neutron and the deuteron, m is the nucleon mass and ε is the binding energy of the deuteron. The solutions of this system of equations are

$$v_p = v_n + 2i\sqrt{\varepsilon/m} \ , \ v_d = v_n + i\sqrt{\varepsilon/m} \tag{9.97}$$

and

$$\frac{Ze^2}{r_0} = E - mv_n^2 - 2i\sqrt{\varepsilon m}v_n + \varepsilon \ , \tag{9.98}$$

where Ze is the ion charge, r_0 is the disintegration point and v_n is real. The mechanical action in equation (9.86) is computed by using $E = p_d^2/m + Ze^2/r$ and $E_p = p_p^2/2m + Ze^2/r$, where E is the energy of the deuteron, E_p is the energy of the proton and $p_{p,n}$ are the momenta of the proton and the neutron. We get the transition amplitude as a function of E and E_p(or E_n); its maximum is achieved for $E_n = 0$.

9.11 Adiabatic transitions

In general, in order to be entitled to apply the time-independent perturbation theory, we need to have stationary states, or linear combinations of stationary states; in the tunneling process described above

[21] E. M. Lifshitz, "The collisions of deuterons with heavy nuclei. I", Phys. Z. Sowjet. (Sov. Phys.) **13** 224 (1938); "The collisions of deuterons with heavy nuclei. II", J. Phys. **1** 199 (1939); "Neutron transfer in collisions of heavy nuclei", ZhETF **9** 237 (1939) (in *Perspectives in Theoretical Physics, The Collected Papers of E. M. Lifshitz*, ed. L. P. Pitaevskii, Pergamon Press (1992)).

we use such stationary states, for static fields (potential barrier). Usually, the interaction is introduced in time; for instance, there is a time for the particle to penetrate the barrier. Indeed, the "attempt" time to approach the barrier can be estimated as $\tau \simeq \hbar/E$, where E is the particle energy (it is the time needed to prepare the state with energy E); therefore, the rate of transmission through barrier is w/τ, where the transmission probability is of the form $w = e^{-\gamma}$, as given above; we can see that the penetration time is τ/w, which is also the lifetime t_{lf} of the particle; if we have a collection of N_0 particles they decay (penetrate the barrier) in time as $N_0 e^{-t/t_{lf}}$. The time τ/w must be long, in comparison with the relevant times, such as to preserve the stationary states; in particular it must be much longer than time τ, *i.e.* we must have $w \ll 1$, as we know we have. We say that the interaction is introduced adiabatically in this case. We may see that only a reasonably weak interaction can be introduced adiabatically, which again is compatible with the assumptions of the perturbation theory.

Similarly, we may introduce adiabatically a (weak) time-dependent interaction. But this is of not much help, because a time-dependent interaction mixes up the states; it generates transitions between the states; we must work then with time-dependent wavefunctions; though, nevertheless, if the strength of the interaction is small, its effects are small. However, we may imagine a situation where the interaction depends very slowly on time, for instance it oscillates very slowly; in the limit of adiabatic variation, the states are not changed; but we may imagine that for slowly varying interaction, the states are slightly perturbed, the energy levels depend slightly on the time; and we may formulate the problem of the transition probability from the state $\psi_1 e^{-\frac{i}{\hbar}E_1 t}$ for $t \rightarrow -\infty$ to the state $\psi_2 e^{-\frac{i}{\hbar}E_2 t}$ for $t \rightarrow \infty$. This is called the problem of adiabatic transitions.[22] It is easy to see that there must be a time t_0, in general complex, when $E_1(t_0) = E_2(t_0)$; and, since $dS = -Edt + pdq$, we get the transition amplitude

$$T = e^{\frac{1}{\hbar} Im \int_c E(t)dt} = e^{-\frac{1}{\hbar} Im \int^{t_0} [E_2(t) - E_1(t)]dt} \quad , \tag{9.99}$$

where the contour C encircles the point t_0 and takes us from E_1 to E_2 (*e.g.*, if $E_2 > E_1$, t_0 is in the upper complex half-plane); the real

[22]See Landau and Lifshitz quoted above.

lower limit of integration is irrelevant. An example of such a transition is the atom ionization under the action of a weak, slowly oscillating electric field,[23] which will be described shortly. This method is called imaginary-time tunneling. Equation (9.99) corresponds to equation (9.73).

It is worth noting a model of imaginary-time tunneling (due to Landau, see the previous footnote). If the transition is going to take place, then it is governed by two wavefunctions $\phi_{1,2}$ with the properties

$$\int dq\phi_1^2 = \int dq\phi_2^2 = 0 \ , \ \int dq\phi_1\phi_2 = 1 \ ; \qquad (9.100)$$

then the combination $\psi = \alpha\phi_! + \beta\phi_2$ gives by $H\psi = E\psi$ the equations

$$\alpha H_{11} + \beta H_{12} = E\beta \ , \ \alpha H_{12} + \beta H_{22} = E\alpha \qquad (9.101)$$

and $E_{1,2} = H_{12} \pm \sqrt{H_{11}H_{22}}$ (where H_{ij}, i, $j = 1, 2$, are the matrix elements of the hamiltonian); if $E_1 = E_2$ at t_0, at least H_{11} or H_{22} must vanish, usually as $t - t_0$; then $E_{1,2} = H_{12} \pm const\sqrt{t - t_0}$ and the energy has two branches for complex time, which contribute to the exponent integral in equation (9.99); the conditions given by equation (9.100) ensure the transition.

The transition amplitudes derived above (including tunneling, reflection above the barrier, dissociation, or imaginary-time tunneling) pertain to two distinct dynamics governed by two distinct hamiltonians. They share the common feature of the quasi-classical dynamics. In quasi-classical dynamics the wavefunction is $e^{\frac{i}{\hbar}S}$, where S is the classical action; the two distinct states have distinct actions $S_{1,2}$, differentiated by two distinct energies $E_{1,2}$, which may depend on the time. The two actions depend on the same coordinate q, which is common for the two states, and time t (also the same, of course). The probability amplitude of transition is $e^{\frac{i}{\hbar}\Delta S}$, where $\Delta S = S_1 - S_2$. The quantum-mechanical states are spread over all the available (common) coordinates, and the transition proceeds for all, indistinctely. But the transition proceeds in time, for quasi-classical motion in long times,

[23] L. V. Keldysh, "Ionization in the field of a strong electromagnetic wave", Sov. Phys.-JETP **20** 1307 (1965) (ZhETF **47** 1945 (1965)).

and, consequently, we should superpose the amplitudes, according to

$$\frac{1}{T}\int dte^{\frac{i}{\hbar}\Delta S} \ , \tag{9.102}$$

where T is the long time of transition. We note that for the same dynamics one of the wavefunctions in equation (9.102) includes, as a coefficient, the matrix elements of the interaction and its time dependence, say, $e^{-i\omega t}$, such that the exponential in equation (9.102) is replaced by a quantity proportional to $e^{\frac{i}{\hbar}(E_2-E_1)t-i\omega t}$; the time integration leads to the well-known transition amplitudes given by the perturbation theory for a time-dependent interaction. For distinct dynamics and quasi-classical wavefunctions the exponential in equation (9.102) varies rapidly in time, except for the saddle points (stationary phase); therefore, we may write

$$\Delta S = \Delta S_0 - (t-t_0)(E_1-E_2) + ... \ , \tag{9.103}$$

and look for the points where

$$E_1 = E_2 \ ; \tag{9.104}$$

for those points, the transition amplitude is $e^{\frac{i}{\hbar}\Delta S_0}$, by the steepest descent (stationary phase) approximation. If $E_1 = E$ and $E_2 = V(x)$, as for a potential barrier, we can see that we should limit ourselves to the region extending from (real) a to b with the notations used above, where $p = p_1 = i\sqrt{2m[V(x)-E]}$ such that $\Delta S_0 = i\int_a^b dx\sqrt{2m[V(x)-E]}$, which gives the tunneling amplitude derived above. For $E_1 = E$ and $E_2 = -kx^2$, as in the example discussed above for the reflection above the barrier, equation (9.104) gives a complex point $x_0 = i\sqrt{\mathcal{E}/k}$ and $\Delta S_0 = \int^{x_0} dx(p_1 - p_2) = 2\int^{x_0} p_1 = 2\int^{x_0}\sqrt{2m(E+kx^2)}$, which leads to the amplitude of reflection above the barrier (the lower bound of integration in ΔS_0 is immaterial). Finally, for a complex-time root t_0 of equation (9.104), we have $\Delta S_0 = -\int^{t_0} dt(E_1 - E_2)$, which leads to the transition amplitude given by equation (9.99).

Let us consider a charge q with mass m in a potential well with the binding energy $-E_b$ ($E_b > 0$). Let us introduce adiabatically an oscillating electric field $\mathcal{E} = \mathcal{E}_0\cos\omega t$, with a low frequency ω, *i.e.* an adiabatic interaction; specifically, we require $\hbar\omega \ll E_b$. The field may derive from a vector potential $A = -(c\mathcal{E}_0/\omega)\sin\omega t$. We assume that $\mid q \mid A_0/mc^2 \ll 1$, where $A_0 = c\mathcal{E}_0/\omega$, *i.e.* $\mathcal{E}_, \ll 10^8 esu$ for $\omega = 10^{15}s^{-1}$, such that the motion remains non-relativistic. In addition, we assume that the amplitude of the charge displacement $\mid q \mid \mathcal{E}_0/m\omega^2$ is much smaller than the dimension a of the bound state, such that the charge be not removed quickly from the potential well; for electrons and atomic-like assemblies $a = 10^{-8}cm$, and we get $\mathcal{E}_0 < 10^4 esu$ (for $\omega = 10^{15}s$; corresponding to a laser intensity $I = 10^{11}w/cm^2$).

The hamiltonian of the charge is

$$H = \frac{1}{2m}\left(\mathbf{p} - \frac{q}{c}\mathbf{A}\right)^2 = \frac{1}{2m}\left(\mathbf{p} + \frac{q\mathbf{E}_0}{\omega}\sin\omega t\right)^2 \quad , \tag{9.105}$$

where $\mathbf{p}$ is the charge momentum. We consider transitions from the state with energy $E_1 = -E_b$ to the state with the energy $E_2 = \frac{q^2\mathcal{E}_0^2}{2m\omega^2}\sin^2\omega t$ $(\mathbf{p} = 0)$;[24] the transition time t_0 is given by

$$\frac{q^2\mathcal{E}_0^2}{2m\omega^2}\sin^2\omega t_0 = -E_b \quad , \tag{9.106}$$

which gives a purely imaginary time $t_0 = i\tau_0$,

$$\sinh\omega\tau_0 = \frac{\omega\sqrt{2mE_b}}{\mid q \mid \mathcal{E}_0} \quad ; \tag{9.107}$$

or,

$$\tau_0 \simeq \frac{1}{\omega}\ln\frac{2\omega\sqrt{2mE_b}}{\mid q \mid \mathcal{E}_0} \quad , \tag{9.108}$$

since $2\omega\sqrt{2mE_b}/ \mid q \mid \mathcal{E}_0 \gg 1$ for our problem. The amplitude of transition probability is given by

$$T = e^{-\frac{1}{\hbar}Im\int^{t_0}dt(E_2 - E_1)} \quad , \tag{9.109}$$

according to the imaginary-time tunneling theory; we get

$$T \simeq e^{-\frac{\mathcal{E}_b}{\hbar\omega}\ln\frac{2\omega\sqrt{2mE_b}}{\mid q\mid\mathcal{E}_0}} \quad ; \tag{9.110}$$

[24] L. V. Keldysh, "Ionization in the field of a strong electromagnetic wave", Sov. Phys.-JETP **20** 1307 (1965) (ZhETF **47** 1945 (1965)).

this is a very small transition amplitude, as it should be. We can see that the criterion $\mid q \mid \mathcal{E}_0 a \ll ma^2\omega^2$ ensures $\omega\sqrt{mE_b} \gg \mid q \mid \mathcal{E}_0$ (and $T \ll 1$). If we extend (unwarrantly) this result to high fields, we get $T \sim e^{-const/\mathcal{E}_0}$. The result is also valid for the ionization of the heavy atoms, ions, molecules, atomic clusters, or ion emission from bound states.[25]

There exists an interesting analogy between the one-dimensional harmonic oscillator and the one-dimensional Schrodinger equation.[26] Indeed, let us consider the one-dimensional Schrodinger equation for a particle with mass m in the potential $V(x)$,

$$\psi'' + \frac{2m}{\hbar^2}[E - V(x)]\psi = 0 \; ; \tag{9.111}$$

by the substitutions $x \leftarrow t$, $\psi \rightarrow x$ and $\sqrt{2m[E - V(x)]}/\hbar = k(x) \rightarrow \omega(t)$, equation (9.111) becomes the equation of motion of a harmonic oscillator

$$\ddot{x} + \omega^2(t)x = 0 \; ; \tag{9.112}$$

we assume that $\omega(t) \rightarrow \omega_1$ for $t \rightarrow -\infty$ and $\omega(t) \rightarrow \omega_2$ for $t \rightarrow +\infty$, $\omega_2 > \omega_1$; for $E > V(+\infty)$ we have a problem of reflection above the barrier. Therefore, we assume the asymptotic forms of the solutions

$$\begin{aligned} x &= A_1 e^{i\omega_1 t} + A_1^* e^{-i\omega_1 t} \; , \; t \rightarrow -\infty \\ x &= A_2 e^{i\omega_2 t} + A_2^* e^{-i\omega_2 t} \; , \; t \rightarrow \infty \; , \end{aligned} \tag{9.113}$$

as in equation (9.88), where $A_2 = \alpha A_1 + \beta A_1^*$. The adiabatic invariant of the harmonic oscillator is $I = E/\omega$, where $E = m\omega^2\overline{x^2}$ is the energy; therefore, we get $I_1 = 2m\omega_1 \mid A_1 \mid^2$ and $I_2 = 2m\omega_2 \mid A_2 \mid^2$, or

$$I_2 = 2m\omega_2[(\mid \alpha \mid^2 + \mid \beta \mid^2) \mid A_1 \mid^2 + 2Re(\alpha\beta^* A_1^2)] \; ; \tag{9.114}$$

---

[25] See also A. M. Perelomov, V. S. Popov and M. V. Terentiev, "Ionization of atoms in an alternating electric field", Sov. Phys.-JETP **23** 924 (1966) (ZhETF **50** 1393 (1966)) and M. V. Ammosov, N. B. Delone and V. P. Krainov, "Tunnel ionization of complex atoms and of atomic ions in an alternating electromagnetic field", Sov. Phys.-JETP **64** 1191 (1986) (ZhETF **91** 2008 (1986)).

[26] A. M. Dykhne, "Quantum transitions in the adiabatic approximation", Sov. Phys.-JETP **11** 411 (1960) (ZhETF **38** 570 (1960)).

it follows, making use of $|\alpha|^2 - |\beta|^2 = \omega_1/\omega_2$ (equation (9.91)),

$$I_2 - I_1 = 4m\omega_2[|\beta|^2|A_1|^2 + Re(\alpha\beta^* A_1^2) \ . \qquad (9.115)$$

The reflection coefficient is given by

$$|R| = |\beta/\alpha| \simeq e^{-2Im\int^{t_0}\omega(t)dt} \ , \qquad (9.116)$$

where t_0 is the solution of the equation $\omega_1 \simeq \omega_2$, as in equation (9.87); for an adiabatic function $\omega(t)$ the reflection coefficient is exponentially small, so is β, and $\alpha \simeq 1$; it follows that the variation of the adiabatic invariant is exponentially small (proportional to β). The equation $\omega_1 = \omega_2$ may be viewed as the equation $e^x = 1 - e^{-x}$, where we take e^x for $x \to -\infty$ and $1 - e^{-x}$ for $x \to +\infty$; the root is $x = i\pi/3$.

9.12 Quasi-classical matrix elements

The quasi-classical wavefunctions of a particle with mass m for two distinct energies $E_{1,2}$ are

$$\psi_1 \simeq e^{\frac{i}{\hbar}\int^x dx\sqrt{2m[E_1 - V(x)]}} \ , \ \psi_2 \simeq e^{\frac{i}{\hbar}\int^x dx\sqrt{2m[E_2 - V(x)]}} \ ; \qquad (9.117)$$

the matrix element of a quantity f is

$$f_{12} = \int dx f e^{\frac{i}{\hbar}[\int^x dx\sqrt{2m[E_1 - V(x)]} - \frac{i}{\hbar}\int^x dx\sqrt{2m[E_2 - V(x)]}]} \qquad (9.118)$$

(where the integration variables x are distinct). For $E_{1,2} > V(x)$ the integrand is rapidly varying; the main contribution comes from the inaccessible region of the motion, where the main contribution comes from the singular points of the potential; we take the singular point x_0 in the complex upper half-plane which gives the greatest contribution to f_{12}; with exponential accuracy we get[27]

$$f_{12} \simeq e^{-\frac{1}{\hbar}\int^{x_0} dx[\sqrt{2m(V - E_1)} - \sqrt{2m(V - E_2)}]} =$$
$$= e^{-\frac{1}{\hbar}Im\int^{x_0} dx[\sqrt{2m(E_1 - V)} - \sqrt{2m(E_2 - V)}]} \ , \qquad (9.119)$$

[27] L. Landau, "Zur Theorie der Energieubertragung bei Stossen", Phys. Z. Sowjet. 1 88 (1932).

where $E_1 < E_2$. With $E_1 = E - \frac{1}{2}\hbar\omega_{21}$, $E_2 = E + \frac{1}{2}\hbar\omega_{21}$, $\hbar\omega_{21} = E_2 - E_1$, we get

$$f_{12} \simeq e^{-\omega_{21}\int^{x_0} dx\sqrt{\frac{m}{2(V-E)}}} = e^{-\omega_{21} Im \int^{x_0} dx\sqrt{\frac{m}{2(E-V)}}} , \tag{9.120}$$

or $f_{12} \simeq e^{-\omega_{21} Im\tau}$, where

$$\tau = \int^{x_0} dx \sqrt{\frac{m}{2(E-V)}} = \int^{x_0} dx \cdot \frac{1}{v} , \tag{9.121}$$

v being the velociy of the particle; this relation is in agreement with the quasi-classical matrix elements as the Fourier transforms. It may happen that the potential V depends on the states 1, 2; then, we should consider among the singular points the roots of the equation $V_1 - E_1 = V_2 - E_2$. Equations (9.119) and (9.120) remain valid for a finite motion.

Let us consider a potential of the form $V = V_0 e^{-\alpha x}$; it has a singular point $x = -\infty$, which we write as $x_0 = \infty e^{i\pi}$; in equation (9.119) we make the substitution $e^{-\alpha x} = y$; the main contribution comes from $\ln x_0$, whose imaginary part is π; the result is

$$f_{12} \simeq e^{-(\pi m/\hbar\alpha)(v_2 - v_1)} , \tag{9.122}$$

where $v_{1,2} = \sqrt{2mE_{1,2}}$ are the asymptotic velocities. For a Coulomb potential and angular momentum zero $f_{12} \simeq e^{-(\pi\alpha/\hbar)(1/v_1 - 1/v_2)}$ (see equation (9.75)). Similarly, for a harmonic potential $V = m\omega^2 x^2/2$, bounded by $V_0/2$ and for $E_{1,2} \gg \hbar\omega$, the singular points are $\pm\infty$ and we get $f_{12} \simeq e^{-\frac{\Delta E}{2\hbar\omega}\ln(V_0/E)}$, where $E_{1,2}$ differ slightly from E.

9.13 Concluding remarks

In the world of the atomic particles we deal with extremely small physical quantities like distances, time durations, energies, momenta, mechanical action, etc; but not necessarily with small velocities; the velocities may be high. For instance, an electron with energy $E = 1eV$, *i.e.* $E = 1.6 \times 10^{-12} erg$, which is a very small energy in comparison with the energies at the macroscopic scale, has a velocity $v = \sqrt{2E/m} \simeq 6 \times 10^7 cm/s$, which is a high velocity in comparison with the macroscopic world (electron mass $m = 10^{-27}g$). The variation of the mechanical action of this electron in time Δt is $\Delta S = E\Delta t$; for $\Delta t \gg 10^{-15}s$ this variation may be much larger than the Planck constant $\hbar = 10^{-27}erg \cdot s$. Similarly, the momentum of this electron is $p = \sqrt{2mE} \simeq 6 \times 10^{-20}g \cdot cm/s$ and the variation of the mechanical action $\Delta S = p\Delta x$ may be much larger than the Planck constant for distances $\Delta x \gg 10^{-8}cm$. The time $10^{-15}s$ is a short time, and the distance $10^{-8}cm$ is a short distance; they are characteristic for the atomic level. It follows that at the atomic level we may have variations ΔS of the mechanical action much larger than the Planck's constant. In this case the wavefunction of the particle has the form $e^{\frac{i}{\hbar}S}$ and the Schrodinger equation reduces to the Hamilton-Jacobi equation for the mechanical action S. This is the quasi-classical behaviour. It means that the particle moves classically, but we may associate a wavefunction of the particle and compute its probability of localization by means of this wavefunction. Of course, since the classical mechanical action is a real function, the localization probability is unity, as for a classical particle. If we allow the classical action to acquire complex values, *i.e.* if we allow the particle to move into classically inaccessible regions, either for real, or complex distances and times, we may get a quantum-mechanical behaviour, since the wavefunction $e^{\frac{i}{\hbar}S}$ acquires an amplitude different from unity in that case, as in the quantum-mechanical case; we get a tunneling in that case. It turns out that this amplitude (tunneling probability) is exponentially small, as expected. We may neglect its amount; its existence serves only to show qualitatively that transitions like disintegration, ionization, etc may happen at the sub-microscopic level. On the other hand, since the quasi-classical wavefunction $e^{\frac{i}{\hbar}S}$ (with real S) should be a one-valued function over a closed orbit of the motion, it provides the quantization rules for slightly-separated (dense) quantum-mechanical states. The existence of the wavefunction is sufficient to show the discrete nature of the quantum-mechanical behaviour for the so-called periodic motions; for aperiodic motions like the tunneling it provides essentially the same information of the quantum-mechanical type, by extending the mechanical action to complex values. We can see that the quasi-classical "approximation" is, in fact, the construction of the whole Quantum Mechanics.

Since the atomic particles are numerous and move with high velocities, we may measure, experimentally, their behaviour, even indirectly; in addition, there exist sufficiently sensitive devices, like the photographic plate, or the bubble chamber, able to detect such particles. For instance, an electric current $I = 1mA = 10^{-3}A$ (*ampere*), where $1A = 3 \times 10^9 esu$ (electron charge is $4.5 \times 10^{-10} esu = 1.6 \times 10^{19} C$, $C = coulomb$, such that $1esu = [1/(3 \times 10^9)]C$) can be measured; it is written as $I = qvnS$, where the charge $q = 4.5 \times 10^{-10} esu$ for electron, v is the electron velocity, n is the density of the electrons and S is the cross-section; with $v = 6 \times 10^7 cm/s$ (energy $1eV$) and an area $S = 1mm^2 = 10^{-2} cm^2$ we get $n \simeq 10^{10} cm^{-3}$. In solid metals the density of the electrons is $n \simeq 10^{22} cm^{-3}$ (Avogadro's number is 6×10^{23}) and, although 10^{10} is much smaller than 10^{22}, we can see that we are able to estimate it (indirectly) from measurements. Therefore, even if the tunneling probabilities are exponentially small, they may be indirectly estimated by experiments.

10 Scattering

10.1 Born scattering

We consider a non-relativistic particle with mass m scattered by a potential $V(t, \mathbf{r})$, where t denotes the time and $\mathbf{r}$ denotes the position; this situation may correspond to the collision of two particles interacting by the potential $V(t, \mathbf{r})$, the scattering being the relative motion of the two particles. The connection between the scattering angles $\theta_{1,2}$ of the two particles and the scattering angle θ of the relative motion is $\tan\theta_1 = m_2 \sin\theta/(m_1 + m_2\cos\theta)$, $\theta_2 = (\pi - \theta)/2$, where $m_{1,2}$ are the masses of the two particles; initially, the particle 2 is at rest.[1]

Making use of the notation $H_0 = p^2/2m$, where $\mathbf{p}$ is the momentum of the particle, the wavefunction ψ_i of the incident particle satisfies the Schrodinger equation

$$i\hbar\frac{\partial\psi_i}{\partial t} = (H_0 + V)\psi_i \tag{10.1}$$

with the initial condition (incoming state)

$$\psi_i^0 = \frac{1}{\sqrt{v}}e^{-\frac{i}{\hbar}E_i t + \frac{i}{\hbar}\mathbf{p}_i\mathbf{r}} \ , \tag{10.2}$$

where $\mathbf{p}_i$ is the initial momentum, $E_i = p_i^2/2m$ is the initial energy and v denotes the volume ($\hbar$ is Planck's constant); the solution of equation (10.1) is given by

$$\psi_i = \psi_i^0 - \frac{i}{\hbar}e^{-\frac{i}{\hbar}H_0 t}\int_{-\infty}^{t} dt' e^{\frac{i}{\hbar}H_0 t'} V\psi_i \ ; \tag{10.3}$$

[1]L. Landau and E. Lifshitz, *Course of Theoretical Physics*, vol.1, *Mechanics*, Elsevier (1976).

the second term on the right in equation (10.3) represents the scattered wavefunction. The wavefunction of the final scattering state (outgoing state) is

$$\psi_f = \frac{1}{\sqrt{v}} e^{-\frac{i}{\hbar}E_f t + \frac{i}{\hbar}\mathbf{p}_f \mathbf{r}} \ , \tag{10.4}$$

where $\mathbf{p}_f$ is the final momentum and $E_f = p_f^2/2m$ is the final energy. We note that the incoming and outgoing states are far away from the scattering region, which, rigourously speaking, limit ourselves to short-range potentials; however, we may extend the treatment to potentials falling rapidly at large distances; the Coulomb potential has a special position in this treatment.

The transition (scattering) amplitude (S-matrix)[2] reads

$$a_{fi} = -\frac{i}{\hbar} \int_{-\infty}^{+\infty} dt (\psi_f, V\psi_i) \ . \tag{10.5}$$

It is convenient to introduce time-independent wavefunctions (the so-called interaction representation)

$$\overline{\psi}_i = e^{\frac{i}{\hbar}H_0 t}\psi_i = \overline{\psi}_i^0 - \frac{i}{\hbar}\int_{-\infty}^{t} dt' \widetilde{V}(t)\overline{\psi}_i \tag{10.6}$$

and

$$\overline{\psi}_f = e^{\frac{i}{\hbar}H_0 t}\psi_f = \frac{1}{\sqrt{v}} e^{\frac{i}{\hbar}\mathbf{p}_f \mathbf{r}} \ , \tag{10.7}$$

where $\overline{\psi}_i^0 = e^{\frac{i}{\hbar}H_0 t}\psi_i^0 = \frac{1}{\sqrt{v}} e^{\frac{i}{\hbar}\mathbf{p}_i \mathbf{r}}$ and

$$\widetilde{V}(t) = e^{\frac{i}{\hbar}H_0 t} V e^{-\frac{i}{\hbar}H_0 t} \ ; \tag{10.8}$$

the scattering amplitude becomes

$$a_{fi} = -\frac{i}{\hbar} \int_{-\infty}^{+\infty} dt (\overline{\psi}_f, \widetilde{V}(t)\overline{\psi}_i) \ , \tag{10.9}$$

[2] J. A. Wheeler, "On the mathematical description of light nuclei by the method of resonating group structure", Phys. Rev. **52** 1107 (1937); W. Heisenberg, "Die "beobachtbaren Grossen" in der Theorie der elementare Teilchen", Z. Phys. **120** 513 (1943).

or, by iteration,

$$a_{fi} = \sum_{n=1}^{\infty} \left(-\frac{i}{\hbar}\right)^n \int_{-\infty}^{+\infty} dt_1 \int_{-\infty}^{t_1} dt_2 ... \int_{-\infty}^{t_{n-1}} dt_n \cdot \\ \cdot (\overline{\psi}_f, \widetilde{V}(t_1)\widetilde{V}(t_2)...\widetilde{V}(t_n)\overline{\psi}_i^0) \; ; \tag{10.10}$$

we may divide by $n!$ and preserve at the same time the chronological order of the operators $\widetilde{V}$.

It is worth noting that we impose scattering boundary conditions (incoming and outgoing wavefunctions), which means that initially and finally, *i.e.* for $t \to \pm\infty$, the particle is outside the range of the potential; at the same time, we assume that the interaction lasts an infinite time, *i.e.* it is introduced adiabatically and it is an adiabatic interaction. Moreover, we assume that the extension of the incident wavefunction along the transverse directions with respect to the incidence direction is larger than the range of the potential.

If we are going to keep higher-order terms in the series expansion of the scattering amplitude (perturbation series), then we aim at solving the Schrodinger equation of the incident particle in the potential V with scattering boundary conditions; if we limit ourselves to the first-order of the perturbation series, the uncertainty in energy occurring in the scattering process is higher than the effects of the potential V. The particle spends the time τ in the potential V, which, for short-range potentials is $\tau = a/v \simeq a\sqrt{m/E}$, where a is the range of the potential, v is the velocity of the particle (not to be mistaken for the volume) and E is the energy of the particle; we must have $\hbar/\tau \gg\mid V \mid$, or $V^2 \ll EE_l$, where $E_l = \hbar^2/ma^2$ is the localization energy in the potential range; the most restrictive form of this condition is $\mid V \mid\ll E_l$ (for $E = E_l$). For the Coulomb potential $V = Ze^2/r$ (with both signs) the velocity of the particle is of the order $v = \sqrt{E/m}$ (the particle has not the chance to be localized), such that $\tau = r\sqrt{m/E}$ (except for $r \to 0$) and the condition $\hbar/\tau \gg Ze^2/r$ gives $E \gg Z^2e^2/a_H$, where $a_H = \hbar^2/me^2$ is the Bohr radius of the particle. If we write the Coulomb potential as $V = \alpha/r$, the condition $\hbar/\tau \gg\mid V \mid$ reads $\hbar v \gg\mid \alpha \mid$. These are the conditions of validity of the Born scattering amplitude.[3] We can see that these conditions allow the application of the quasi-classical behaviour.

[3]M. Born, "Quantenmechanik der Stossvorgange", Z. Phys. **38** 803 (1926).

From equation (10.10) the Born amplitude is given by

$$a_{fi} = -\frac{i}{\hbar}\int_{-\infty}^{+\infty} dt(\overline{\psi}_f, \widetilde{V}(t)\overline{\psi}_i^0) =$$
$$= -\frac{i}{\hbar v}\int_{-\infty}^{+\infty} dt e^{\frac{i}{\hbar}(E_f-E_i)t}\int d\mathbf{r}V(t,\mathbf{r})e^{-\frac{i}{\hbar}(\mathbf{p}_f-\mathbf{p}_i)\mathbf{r}} \; ; \tag{10.11}$$

we denote by $\Delta E = E_f - E_i$ the energy transfer and by $\mathbf{p} = \mathbf{p}_f - \mathbf{p}_i$ the momentum transfer. It is obvious that the Fourier components $e^{-i\omega t}e^{i\mathbf{qr}}$ of the potential give a scattering amplitude

$$a_{fi}(\omega, \mathbf{q}) = -\frac{i\hbar^3}{v}V(\omega, \mathbf{q})\delta(\Delta E - \hbar\omega)\delta(\mathbf{p} - \hbar\mathbf{q}) \; , \tag{10.12}$$

where $V(\omega, \mathbf{q})$ is the Fourier component of the potential; we can see that the scattering takes place by absorbing (emitting) quanta of energy $\hbar\omega$ and quanta of momentum $\hbar\mathbf{q}$ from (to) the potential; these quanta are called virtual particles, because, although they conserve the energy and the momentum, they do not have a relationship between energy and momentum. The scattering amplitude is

$$a_{fi} = \int d\omega d\mathbf{q} a_{fi}(\omega, \mathbf{q}) = -\frac{i}{\hbar v}V(\Delta E/\hbar, \mathbf{p}/\hbar) \; ; \tag{10.13}$$

$\mid a_{fi} \mid^2 d\nu$ is the scattering probability, where $d\nu = vd\mathbf{p}_f/(2\pi\hbar)^3$ is the number of final states, $dw_{fi} =\mid a_{fi} \mid^2 d\nu/\delta t$ is the scattering probability per unit time, where δt is the scattering (collision) time (of the order of τ), and $d\sigma = dw_{fi}/(v_i/v) = v \mid a_{fi} \mid^2 d\nu/v_i\delta t$ is the cross-section, where v_i is the velocity of the incident particle and v_i/v is the current density:

$$d\sigma = \frac{\mid V(\Delta E/\hbar, \mathbf{p}/\hbar) \mid^2}{\hbar^2 v_i \delta t}\frac{d\mathbf{p}_f}{(2\pi\hbar)^3} \; . \tag{10.14}$$

If the potential is periodic in time, *i.e.* if it is of the form $V(t, \mathbf{r}) = e^{-i\omega_0 t}V(\mathbf{r}) + c.c.$, then its Fourier transform is $V(\omega, \mathbf{q}) = 2\pi[\delta(\omega - \omega_0) + \delta(\omega + \omega_0)]V(\mathbf{q})$; the cross-section for an energy exchange $\hbar\omega_0$ is

$$d\sigma = \frac{\mid V(\mathbf{p}/\hbar) \mid^2}{2\pi\hbar^3 v_i \delta t}\delta^2(\Delta E - \hbar\omega_0)d\mathbf{p}_f \; ; \tag{10.15}$$

we write one of the delta function in equation (10.15) as $\delta(\Delta E-\hbar\omega_0)=1/\delta E$ and replace $\delta t\delta E$ by $2\pi\hbar$; we get

$$d\sigma=\frac{\mid V(\mathbf{p}/\hbar)\mid^2}{(2\pi\hbar^2)^2 v_i}\delta(\Delta E-\hbar\omega_0)d\mathbf{p}_f \ ; \tag{10.16}$$

we replace in this equation $d\mathbf{p}_f$ by $p_f^2dp_fd\Omega$, where $d\Omega$ is the infinitesimal solid angle with respect to $\mathbf{p}_i$ and integrate over p_f; the cross-section becomes

$$d\sigma=\frac{v_f}{v_i}\left|\frac{m}{2\pi\hbar^2}\int d\mathbf{r}V(\mathbf{r})e^{-\frac{i}{\hbar}\mathbf{pr}}\right|^2 d\Omega \ , \tag{10.17}$$

where $\mathbf{p}=\mathbf{p}_f-\mathbf{p}_i$ and $p_f=\sqrt{p_i^2+2m\hbar\omega_0}$; $d\Omega=\sin\theta d\theta d\varphi$, where θ, φ are the spherical angles made by $\mathbf{p}_f$ with $\mathbf{p}_i$ (scattering angles). If $\omega_0=0$ (potential independent of time), the collision is elastic and $p=2p_i\sin\theta/2$ ($v_f=v_i$). It is convenient to write the elastic Born scattering cross-section as

$$d\sigma=\mid f(\theta)\mid^2 d\Omega \ , \ f(\theta)=-\frac{m}{2\pi\hbar^2}\int d\mathbf{r}V(\mathbf{r})e^{-i\mathbf{qr}} \ , \tag{10.18}$$

where $\mathbf{q}=\mathbf{k}'-\mathbf{k}$, ($\mathbf{p}_i=\hbar\mathbf{k}$, $\mathbf{p}_f=\hbar\mathbf{k}'$) $q^2=2k^2(1-\cos\theta)$, θ being the scattering angle. For the Coulomb potential $V=\alpha/r$ we get $d\sigma=4m^2\alpha^2/p^4=\alpha^2/16E_i^2\sin^4\theta/2$. This is the Rutherford formula for classical scattering.[4]

10.2 Scattering amplitude, phase shifts

Let us consider an elastic collision of a particle with mass m in a centrally symmetric potential $V(r)$. We leave aside the time factors in the wavefunctions and the normalization constants; the wavefunction should satisfy the Schrodinger equation

$$\Delta\psi+k^2\psi=\frac{2m}{\hbar^2}V \ , \tag{10.19}$$

[4]E. Rutherford, "The scattering of α and β particles by matter and the structure of the atom", Phil. Mag. **21** 669 (1911); N. F. Mott, "The solution of the wave equation for the scattering of particles by a coulombian centre of force", Proc. Roy. Soc. **A118** 542 (1928); W. Gordon, "Uber den Stoss zweier Punktladungen nach der Wellenmechanik", Z. Phys. **48** 180 (1928).

where $k^2 = 2mE/\hbar^2$, E being the energy of the particle. In the first order of the perturbation theory we seek the solution as $\psi = e^{i\mathbf{kr}} + \psi_s$, where the scattered wave ψ_s obeys the equation $\Delta\psi_s + k^2\psi_s = (2m/\hbar^2)Ve^{i\mathbf{kr}}$; we get the solution

$$\psi = e^{i\mathbf{kr}} - \frac{m}{2\pi\hbar^2}\int d\mathbf{r}' V(\mathbf{r}')e^{i\mathbf{kr}'}\frac{e^{ik|\mathbf{r}-\mathbf{r}'|}}{|\mathbf{r}-\mathbf{r}'|} \ , \tag{10.20}$$

where we retain only the outgoing waves (retarded Kirchhoff's potential); at large distances we have

$$\psi \simeq e^{i\mathbf{kr}} - \frac{m}{2\pi\hbar^2}\int d\mathbf{r}' V(\mathbf{r}')e^{i(\mathbf{k}-k\mathbf{r}/r)\mathbf{r}'}\frac{e^{ikr}}{r} \ , \tag{10.21}$$

where $\mathbf{k}' = k\mathbf{r}/r$ is the wavevector of the final state; therefore, we may write the asymptotic wavefunction as

$$\psi \simeq e^{i\mathbf{kr}} + f(\theta)\frac{e^{ikr}}{r} \ , \tag{10.22}$$

where

$$f(\theta) = -\frac{m}{2\pi\hbar^2}\int d\mathbf{r}' V(\mathbf{r}')e^{-i(\mathbf{k}'-\mathbf{k})\mathbf{r}'} \ ; \tag{10.23}$$

θ is the angle made by the vectors $\mathbf{k}$ and $\mathbf{k}'$. The probability per unit time that the scattered particle will pass through the surface $dS = r^2 d\Omega$ is $(|f|^2/r^2)vdS = |f|^2 vd\Omega$, where v is the velocity of the particle and $d\Omega$ is the solid angle; it follows that the cross-section is $d\sigma = |f|^2 d\Omega$, as in equation (10.18); f is the (Born) scattering amplitude.

The asymptotic form of the scattering wavefunction given by equation (10.22) is general (for centrally symmetric potentials); if we expand $f(\theta)$ in Legendre polynomials $P_l(\cos\theta)$, we get

$$f(\theta) = \sum_l (2l+1)f_l P_l(\cos\theta) \ , \tag{10.24}$$

where the coefficients f_l are called partial-wave amplitudes; the expansion given by equation (10.24) is called the partial-wave expansion of the scattering amplitude.

The general solution of the Schrodinger equation with a centrally symmetric potential can be written as

$$\psi = \sum_l A_l P_l(\cos\theta) R_{kl}(r) \ , \tag{10.25}$$

where $P_l(\cos\theta)$ are the Legendre polynomials, A_l are undetermined coefficients and the radial function R_{kl} satisfies the equation

$$\frac{1}{r^2}\frac{d}{dr}\left(r^2\frac{dR_{kl}}{dr}\right) + \left[k^2 - \frac{l(l+1)}{r^2} - \frac{2m}{\hbar^2}V\right] R_{kl} = 0 \ ; \tag{10.26}$$

at large distances we may neglect both the centrifugal potential $l(l+1)/r^2$ and the short-range potential $V(r)$; since $(1/r^2)\frac{d}{dr}(r^2 dR_{kl}/dr)$ can be written as $(1/r)\frac{d^2}{dr^2}(rR_{kl})$, we have the asymptotic equation

$$\frac{1}{r}\frac{d^2}{dr^2}(rR_{kl}) + k^2 R_{kl} = 0 \ , \tag{10.27}$$

whose solution is

$$R_{kl} = \frac{2}{r}\sin(kr - l\pi/2 + \delta_l) \ ; \tag{10.28}$$

the phase shifts δ_l (in the range 0 to π) account for the potential; we may write the asymptotic wavefunction as

$$\psi \simeq -i\sum_l A_l P_l(\cos\theta)\left[e^{-il\pi/2}e^{i\delta_l}\frac{e^{ikr}}{r} - e^{il\pi/2}e^{-i\delta_l}\frac{e^{-ikr}}{r}\right] \ . \tag{10.29}$$

On the other hand, the expansion of the plane wave in spherical Bessel functions j_l is

$$e^{i\mathbf{kr}} = \sum_l i^l(2l+1)P_l(\cos\theta)j_l(kr) \tag{10.30}$$

(where θ is the angle between $\mathbf{k}$ and $\mathbf{r}$) with the asymptotic behaviour

$$e^{i\mathbf{kr}} \simeq -\frac{i}{2k}\sum_l(2l+1)P_l(\cos\theta)\left[\frac{e^{ikr}}{r} - (-1)^l\frac{e^{-ikr}}{r}\right] \ ; \tag{10.31}$$

if we choose $A_l = [(2l+1)/2k]i^l e^{i\delta_l}$, the contributions of the incoming spherical waves to ψ and the plane wave $e^{i\mathbf{kr}}$ are the same and the contribution of the outgoing spherical waves to ψ

$$\begin{gathered} -i\sum_l A_l P_l(\cos\theta)(-i)^l e^{i\delta_l}\frac{e^{ikr}}{r} = \\ = -i\sum_l \frac{2l+1}{2k}P_l(\cos\theta)e^{2i\delta_l}\frac{e^{ikr}}{r} \end{gathered} \tag{10.32}$$

should be equal to the contribution

$$\left[-i\sum_l \frac{2l+1}{2k}P_l(\cos\theta) + f\right]\frac{e^{ikr}}{r} \tag{10.33}$$

of the outgoing spherical waves to the right side $e^{i\mathbf{kr}} + fe^{ikr}/r$ of equation (10.22); hence, we get

$$f(\theta) = \frac{i}{2k}\sum_l (2l+1)P_l(\cos\theta)\left(1 - e^{2i\delta_l}\right) \quad , \tag{10.34}$$

which is the expansion of the scattering amplitude in Legendre polynomials, $f_l = i(1 - e^{2i\delta_l})/2k$ being the partial amplitudes.[5] The total cross-section is

$$\begin{gathered} \sigma = \int d\Omega \mid f(\theta) \mid^2 = 4\pi\sum_l (2l+1) \mid f_l \mid^2 = \\ = \frac{4\pi}{k^2}\sum_l (2l+1)\sin^2\delta_l \ . \end{gathered} \tag{10.35}$$

We can see that $4\pi Imf(0) = k\sigma$; this is called the optical theorem for scattering. If we use the expansion

$$e^{i\mathbf{kr}} = 4\pi\sum_{lm} i^l j_l(kr)Y_{lm}(\theta\varphi)Y^*_{lm}(\theta', \varphi') \ , \tag{10.36}$$

where Y_{lm} are the spherical waves, j_l are the spherical Bessel functions, θ, φ give the direction of $\mathbf{k}$ and θ', φ' give the direction of $\mathbf{r}$, then we can put the Born scattering amplitude given by equation

[5] H. Faxen und J. P. Holtsmark, "Beitrag zur Theorie des Durchganges langsamer Elektronen durch Gase", Z. Phys. **45** 307 (1927).

(10.23) for a centrally symmetric potential in the form given by equation (10.34) with small δ_l given by

$$\delta_l = -\frac{\pi m}{\hbar^2}\int dr \cdot rV(r)J^2_{l+1/2}(kr) \tag{10.37}$$

($j_l(z) = \sqrt{\pi/2z}J_{l+1.2}(z)$). In general, the connection between the phase shifts δ_l and the potential $V(r)$ (the so-called inverse problem in the potential theory) can only be obtained by solving the Schrodinger equation. For Coulomb potentials there exist logarithmic phase shifts, which do not contribute to the scattering amplitude.[6] The Coulomb potential represents a special case which will be treated separately below. Also, we note that the Born scattering amplitude does not fulfil the optical theorem.

10.3 Quasi-classical behaviour

We can use equation (10.34) to get the quasi-classical scattering cross-section. According to equation (10.26)

$$p = \hbar\sqrt{k^2 - l(l+1)/r^2 - (2m/\hbar^2)V} \tag{10.38}$$

is the radial momentum within the quasi-classical approximation; it should have large values, k and l should be large and $l(l+1)$ should be replaced by $(l+1/2)^2 \simeq l^2$. The phase shifts are given by

$$\begin{aligned}\delta_l = &\int_{r_0}^{\infty} dr\sqrt{k^2 - l(l+1)/r^2 - (2m/\hbar^2)V} - \\ &- \int_{r_0}^{\infty} dr\sqrt{k^2 - l(l+1)/r^2}\ ,\end{aligned} \tag{10.39}$$

where r_0 is the root of the equation $p = 0$ ($r > r_0$ is the classically accessible region); we can see that $r_0 \simeq l/k$; the phase shifts are given

[6]In general, wavefunctions of the form given by equation (10.25) are used in the continuum spectrum for transitions caused by additional perturbations (A. Sommerfeld, "Uber die Beugung und Bremsung der Elektronen", Ann. Physik **403** 257 (1931); see also, N. F. Mott and H. S. W. Massey, *Theory of Atomic Collisions*, Oxford University Press (1965).

approximately by

$$\delta_l \simeq -\frac{m}{\hbar^2}\int_{r_0}^{\infty} dr \frac{V(r)}{\sqrt{k^2 - l(l+1)/r^2}} \simeq mV(r_0)r_0/\hbar^2 k \ ; \qquad (10.40)$$

in the quasi-classical appoximation the phase shifts should be small, $\delta_l \ll 1$; we have

$$\sigma \sim \frac{1}{k^2}\sum_l (2l+1)\delta_l^2 \sim \int dr_0 V^2(r_0) r_0^2 \ ; \qquad (10.41)$$

if the potential goes like $V \sim 1/r^n$ at infinity, this integral converges for $n > 2$; similar arguments in equation (10.34) show that the forward scattering amplitude $f(\theta = 0)$ goes like $\int dr_0 V(r_0) r_0^2$; it diverges for $n \leq 3$. We note also that the phase shifts given by equation (10.40) diverge for $n \leq 1$ (*e.g.*, for the Coulomb potential). The forward scattering can be omitted in this case, by noticing that equation (10.34) can also be written as

$$f(\theta) = \frac{1}{2ik}\sum_l (2l+1) P_l(\cos\theta) e^{2i\delta_l} \qquad (10.42)$$

for $\theta \neq 0$, since

$$\sum_l (2l+1) P_l(\cos\theta) = 4\delta(1 - \cos\theta) \qquad (10.43)$$

(by the orthogonality of the Legendre polynomials); then, in equation (10.42) we can use $\delta_l - \delta_0$ instead of δ_l, which are finite.

We turn now to the quasi-classical approximation. Let us assume $l\theta \gg 1$; then, the Legendre polynomials go like

$$P_l \simeq \sqrt{\frac{2}{\pi l \sin\theta}} \sin[(l+1/2)\theta + \pi/4] \qquad (10.44)$$

and the main contribution to the scattering amplitude given by equation (10.42) comes from

$$2\frac{d\delta_l}{dl} \pm \theta = 0 \ ; \qquad (10.45)$$

this equation gives a value l_0 as a function of θ and k. The derivative of the phase shifts can be calculated from equation (10.39); with the substitution $\hbar l_0 = mv\rho$, where v is the velocity of the particle and ρ is the classical impact parameter ($k = mv/\hbar$), equation (10.45) becomes

$$\int_{r_0}^{\infty} dr \frac{(mv\rho/r^2)dr}{\sqrt{2m(E-V)-(mv\rho/r)^2}} = \frac{1}{2}(\pi \mp \theta) \ , \tag{10.46}$$

where $E = \hbar^2 k^2/2m$ is the energy of the particle; this is the equation of motion for the classical particle (the signs $\mp$ correspond to repulsive/attractive potentials); it gives l_0 (ρ) as a function of θ and E. Now we compute the scattering amplitude by the method of the stationary phase, where $d^2\delta_l/dl_0^2 = d\theta/2dl_0$ for $l = l_0$; we get

$$f(\theta) \simeq \frac{1}{k}\left(\frac{l_0}{\sin\theta}\frac{dl_0}{d\theta}\right)^{1/2} \tag{10.47}$$

up to an imaginary exponential; hence, the cross-section

$$d\sigma = \mid f(\theta) \mid^2 \cdot 2\pi \sin\theta d\theta = 2\pi(l_0/k^2)dl_0 = 2\pi\rho d\rho \ , \tag{10.48}$$

which is the classical scattering cross-section; it is valid for $\theta \gg 1/l_0 = \hbar/mv\rho$. The derivation given above holds for $\theta \neq 0$; the forward scattering is never quasi-classical (for short-range potentials, where the phase shifts are finite). For slowly-decreasing potentials, like the Coulomb potentials, the forward scattering is quasi-classical. The Coulomb potential represents a special case: due to its slow variation and its long range, the scattering may always be considered quasi-classical in this potential; this explains the coincidence of the Rutherford cross-section computed either by Born scattering or by using partial waves.

If the energy of the particle is sufficiently high, the scattering is practically at small θ, the particle has trajectory (it is quasi-classical) and the momentum transfer $\mathbf{q}$ is practically perpendicular to the wavevector $\mathbf{k}$; we can introduce the longitudinal coordinate z and the transverse coordinate $\mathbf{r}_\perp$; then the Born cross-section can be written as

$$d\sigma = \left| \frac{m}{2\pi\hbar^2} \int d\mathbf{r}_\perp dz V(\sqrt{r_\perp^2 + z^2} e^{-i\mathbf{q}\mathbf{r}_\perp} \right|^2 d\Omega \ ; \tag{10.49}$$

on the other hand, from equation (10.40) the phase shifts are

$$\delta \simeq -\frac{m}{2\hbar^2 k}\int_{-\infty}^{+\infty} dz V(\sqrt{r_\perp^2+z^2} \ll 1 \; ; \tag{10.50}$$

hence,

$$\begin{aligned} d\sigma &\simeq \left|\tfrac{k}{\pi}\int d\mathbf{r}_\perp \delta e^{-i\mathbf{q}\mathbf{r}_\perp}\right|^2 d\Omega = \\ &= \left|\tfrac{k}{2\pi i}\int d\mathbf{r}_\perp \left(e^{2i\delta}-1\right)e^{-i\mathbf{q}\mathbf{r}_\perp}\right|^2 d\Omega \end{aligned} \tag{10.51}$$

and

$$f \simeq \frac{k}{2\pi i}\int d\mathbf{r}_\perp \left(e^{2i\delta}-1\right)e^{-i\mathbf{q}\mathbf{r}_\perp} \; ; \tag{10.52}$$

this equation can be obtained from equation (10.42) by using

$$P_l(\cos\theta) \simeq J_0(\theta l) = \frac{1}{2\pi}\int_0^{2\pi} d\varphi e^{-i\theta l\cos\varphi} \quad , \tag{10.53}$$

where θ is small, l is large and J is the Bessel function of zeroth order ($q = k\theta$, $\rho = l/k$).

Finally, we note that equation (10.46) may have complex roots. Then, the mechanical action is $S(\theta) = mv\int d\theta\rho(\theta)$ and the scattering amplitude is given by $f \simeq e^{-\frac{1}{\hbar}ImS(\theta)}$ with exponential accuracy.

10.4 Luminiscence and rainbow

It may happen that the deviation angle θ is zero for some value $\rho = l_0/k$ of the impact parameter; this defines the luminiscence scattering; we have the expansion

$$\delta_l = \delta_{l_0} + \frac{1}{2}\beta l'^2 \tag{10.54}$$

in powers of $l' = l - l_0$ for the phase shifts; according to equation (10.45), $\theta = 0$ for $l = l_0$; we introduce this expression in equation (10.42) and use equation (10.53) for P_l; we get

$$f = \frac{l_0}{ik}e^{2i\delta_{l_0}}\int dl' J_0(l\theta)e^{i\beta l'^2} \quad , \tag{10.55}$$

where the integration is performed over a range $1/\sqrt{\beta}$; we may take the Bessel function outside the integral, and get the cross-section

$$d\sigma = \frac{\pi l_0^2}{\beta k^2} J_0^2(l_0\theta) d\Omega \ . \tag{10.56}$$

If the scattering angle θ exhibits an extremum θ_0 for some value of the impact parameter $\rho = l_0/k$, then, according to equation (10.45), we may write the series expansion

$$2\delta_l = 2\delta_{l_0} + \theta_0 l' + \frac{1}{3}\alpha l'^3 \ ; \tag{10.57}$$

this is the rainbow scattering; instead of equation (10.47) we have now

$$\mid f(\theta) \mid = \frac{1}{k}\left(\frac{l_0}{2\pi \sin\theta_0}\right)^{1/2} \int dl' e^{i(-l'\theta' + \frac{1}{3}\alpha l'^3)} \ , \tag{10.58}$$

where $\theta' = \theta - \theta_0$; the integral is related to the Airy function Φ; the cross-section is

$$d\sigma = \frac{4\pi l_0}{\alpha^{2/3} k^2} \Phi^2(-\theta'/\alpha^{1/3}) d\theta' \ . \tag{10.59}$$

The cross-section decreases in the classically inaccessible region ($\theta' > 0$ for $\alpha < 0$, or $\theta' < 0$ for $\alpha > 0$) and oscillates on the other side of the point θ_0, with a decreasing amplitude; the maximum is reached for $\theta'/\alpha^{1/3} \simeq 1$.

10.5 Slow particles

If the particles are slow the Born cross-section is

$$d\sigma \simeq \left| \frac{m}{2\pi\hbar^2} \int d\mathbf{r} V(\mathbf{r}) \right|^2 d\Omega \ ; \tag{10.60}$$

it does not depend on the scattering angle (the scattering is in s-wave) (it corresponds formally to a partial-wave amplitude $f = f_0 = \delta_0/k$ (equation (10.34)), *i.e.* $\delta_0 \sim k$). We note that for centrally symmetric potentials the integration over angular variables in the Born amplitude (equation (10.23)) yields a factor $\sin qr/qr$, where $q^2 = 2k^2(1-\cos\theta)$; a formal expansion in powers of qr has terms $\sim k^{2l}(1-\cos\theta)^l$, which are proportional to $k^{2l}P_l(\cos\theta)$ for $k \to 0$; it follows, from equation (10.24), that the partial amplitudes go like $f_l \sim k^{2l}$ and the phase shifts go like $\delta_l \sim k^{2l+1}$ in the limit of low energy, *i.e.* they are vanishing in this limit. In general, from equation (10.37) we can see that the phase shifts are $\delta_l \simeq (2l+3)^{-1}(V/E_l)(ka)^{2l+1}$ for short-range potentials V (potential range a) and $ka \ll 1$, where $E_l = \hbar^2/ma^2$ is the localization energy (also, for higher energy δ_l are small).

The scattering of slow particles exhibits a particularity. It may happen that the validity condition $V^2 \ll EE_l$ of the Born scattering is not fulfilled (for Coulomb potential α/r this condition reads $\mid \alpha \mid \gg \hbar v$, where v is the velocity of the particle). In this case we need to reconsider the derivation of the Born amplitude. The results are valid for short-range potentials.

The Schrodinger equation for a particle with mass m in a potential $V(\mathbf{r})$ reads

$$(\Delta + k^2)\psi = 4\pi U(\mathbf{r})\psi \ , \tag{10.61}$$

where E is the energy of the particle, $k^2 = 2mE/\hbar^2$ and $U(\mathbf{r}) = mV(\mathbf{r})/2\pi\hbar^2$. We may write the solution in the form $\psi(\mathbf{r}) = e^{ikz} + \varphi(\mathbf{r})$, where $\varphi(\mathbf{r})$ behaves as an outgoing wave at infinity; we assume that the incoming wave propagates along the z-axis. As it is well-known from the theory of the potential, this solution reads

$$\varphi(\mathbf{r}) = -\int d\mathbf{r}' \cdot U(\mathbf{r}')[e^{ikz'} + \varphi(\mathbf{r}')]\frac{e^{ik|\mathbf{r}-\mathbf{r}'|}}{|\mathbf{r}-\mathbf{r}'|} \ . \tag{10.62}$$

It is the integral Schrodinger equation, and its *rhs* contains the Green function of the Schrodinger operator. If the potential $U(\mathbf{r})$ decreases conveniently fast at infinity we may disentangle the integration with respect to $\mathbf{r}'$ in equation (10.62) from the $\mathbf{r}$-dependence, by writing $|\mathbf{r}-\mathbf{r}'| \simeq r - \mathbf{r}\mathbf{r}'/r$ for $r \gg r'$, and get the asymptotic form of solution

$$\varphi(\mathbf{r} \sim \infty) = -[U(\mathbf{q}) + \int d\mathbf{r}' \cdot U(\mathbf{r}')\varphi(\mathbf{r}')e^{-i\mathbf{k}'\mathbf{r}'}]\frac{e^{ikr}}{r} \ , \tag{10.63}$$

where

$$U(\mathbf{q}) = \int d\mathbf{r} \cdot U(\mathbf{r})e^{-i\mathbf{q}\mathbf{r}} \tag{10.64}$$

is the Fourier transform of the potential, $\mathbf{k}' = k\mathbf{r}/r$ is the wavevector of the scattered wave ($\mathbf{k}$ being the wavevector of the incoming wave) and $\mathbf{q} = \mathbf{k}' - \mathbf{k}$ is the wavevector transfer, $q = |\mathbf{k}' - \mathbf{k}| = 2k\sin\theta/2$. In equation (10.63) $-U(\mathbf{q})$ is the well-known Born's scattering amplitude.

In order to solve equation (10.62) for $\varphi(\mathbf{r})$ we need the wavefunction $\varphi(\mathbf{r})$ for small values of $\mathbf{r}$, *i.e.* for r comparable with the range of the potential. Let us denote it by $\varphi(\mathbf{r} \sim 0)$ and the range of the potential by b. We can write from equation (10.62)

$$\varphi(\mathbf{r} \sim 0) \simeq -[U(-\mathbf{k}) + \int d\mathbf{r}' \cdot U(\mathbf{r}')\varphi(\mathbf{r}')]\frac{e^{ikb}}{b} \ , \tag{10.65}$$

or

$$b'\varphi(\mathbf{r} \sim 0) \simeq -U(-\mathbf{k}) - \int d\mathbf{r}' \cdot U(\mathbf{r}')\varphi(\mathbf{r}') \ , \tag{10.66}$$

where $b' = be^{-ikb} \simeq b(1 - ikb)$; b'^{-1} is an order-of-magnitude estimation for the factor $e^{ik|\mathbf{r}-\mathbf{r}'|}/|\mathbf{r} - \mathbf{r}'|$. The approximation used in deriving equation (10.66) is valid for weak potentials and $kb \ll 1$. We multiply equation (10.66) by $U(\mathbf{r})$ and integrate to get

$$\begin{gathered} b' \textstyle\int d\mathbf{r} \cdot U(\mathbf{r})\varphi(\mathbf{r}) \simeq \\ \simeq -U(\mathbf{k}=0)U(-\mathbf{k}) - U(\mathbf{k}=0) \textstyle\int d\mathbf{r}' \cdot U(\mathbf{r}')\varphi(\mathbf{r}') \ , \end{gathered} \tag{10.67}$$

whence

$$\int d\mathbf{r} \cdot U(\mathbf{r})\varphi(\mathbf{r}) = -\frac{U(\mathbf{k}=0)U(-\mathbf{k})}{b' + U(\mathbf{k}=0)} \ , \tag{10.68}$$

and

$$\varphi(\mathbf{r} \sim 0) \simeq -\frac{U(-\mathbf{k})}{b' + U(\mathbf{k}=0)} \ . \tag{10.69}$$

The asymptotic solution (10.63) becomes now

$$\psi(\mathbf{r} \sim \infty) = -[U(\mathbf{q}) - \frac{U(-\mathbf{k})U(\mathbf{k}')}{b' + U(\mathbf{k}=0)}]\frac{e^{ikr}}{r} \ . \tag{10.70}$$

It follows that the scattering amplitude can be represented as[7]

$$f = -U(\mathbf{q}) + \frac{U(-\mathbf{k})U(\mathbf{k}')}{b' + U(\mathbf{k}=0)} \ . \tag{10.71}$$

[7]M. Apostol, "Elastic scattering in short-range potentials", Roum. Reps. Phys. **59** 5 (2007).

The general scheme for estimating the scattering amplitude given by equation (10.71) is to notice that $U(\mathbf{k})$ can be represented as $U(\mathbf{k}) \sim b(V/\varepsilon)$, where V is the average interaction and $\varepsilon \sim \hbar^2/mb^2$ is the localization energy over the potential range; then, the scattering amplitude reads $f \sim -b(V/\varepsilon)/(1+V/\varepsilon)$. Although equation (10.71) has a perturbational character, its second term on the right differs from the well-known higher-order terms of the perturbation theory. This tells that the usual expansion of the wavefunction as a perturbation series may not be warranted in general, as the closed, finite result obtained above indicates.

We note that the scattering amplitude is a matrix element with labels $\mathbf{k}$ and $\mathbf{k}^{'}$; the scattering matrix $S(\mathbf{k},\mathbf{k}^{'}) = \delta_{\mathbf{k}\mathbf{k}'} + f(\mathbf{k},\mathbf{k}^{'})$ properly normalized is unitary (conserves the number of particles) and satisfies the "reciprocity theorem" ($f(\mathbf{k},\mathbf{k}') = f(-\mathbf{k}',-\mathbf{k})$, time-reversal symmetry). For centrally symmetric potentials the angular dependence in equation (10.71) comes through $k^2\cos\theta$ in Born's amplitude, so that in an expansion $\sum(2l+1)f_lP_l$ in Legendre polynomials, the partial-wave amplitudes f_l go like $k^{2l} \sim E^{-l}$, where l is the angular momentum.

Equation (10.71) implies a fundamental result for the scattering theory. We arrive at it by making use of $b' = be^{-ikb} \simeq b - ikb^2$. Then, we get straightforwardly that the scattering amplitude f given by equation (10.71) can be written as

$$f \sim \frac{1}{g-ik} \ , \tag{10.72}$$

where g is a real function of $\mathbf{k}$ and $\mathbf{k}'$. According to equation (10.71) it is expressible in terms of the interaction. This is a basic result, which follows from equation (10.63). It shows that f is finite and never singular.[8] Finally, it can be checked on equation (10.71) that the "optical theorem" $Imf(\mathbf{k},\mathbf{k}) = Imf(0) = k\sigma/4\pi$ is satisfied; then, the partial-wave amplitudes are such that $f_l = (g_l - ik)^{-1}$. In the

[8] *i.e.* the scattering amplitude is analytic in energy, for $E > 0$; it may be extended to $E < 0$, where it has discrete poles, corresponding to bound states (which can also be approached with the integral equation of the potential); the origin $E = 0$ is a branch point, and since the imaginary part for $ReE > 0$ cannot be changed, it follows that a cut must be practised on the right half of the energy axis.

long wavelengths limit $kb \ll 1$ the Fourier transform of the potential in equation (10.71) has a weak k-dependence, so we may neglect it;[9] the optical theorem is then satisfied.

10.6 Potential $\delta(\mathbf{r})$

Let us assume a potential $V(\mathbf{r}) = C\delta(\mathbf{r})$, *i.e.* $U(\mathbf{r}) = mC\delta(\mathbf{r})/2\pi\hbar^2 = U_0\delta(\mathbf{r})$. Equation (10.62) has then the exact solution

$$\varphi(r) = -U_0[1+\varphi(0)]\frac{e^{ikr}}{r} . \tag{10.73}$$

This equation has no acceptable solution; for instance, $\varphi(r) = 0$ for any finite r, and $\varphi(0) = -1$. We write $C = Vb^3$ where $b \to 0$ and $V \to \infty$ such that C is finite. Then, we get $\varphi(0) = -U_0/(b'+U_0) \simeq -1 + b'/U_0 + ...$, where $b' = be^{-ikb}$, and

$$\varphi(r) = -\frac{b'U_0}{b'+U_0}\cdot\frac{e^{ikr}}{r} \simeq -b'\frac{e^{ikr}}{r} , \tag{10.74}$$

and the corresponding scattering amplitude $f = -b'U_0/(b'+U_0) \simeq -b'$, which coincides with equation (10.71) for $U(\mathbf{k}) = U_0$ and $b \to 0$. The representation given by equation (10.72) reads

$$f = \frac{1}{-U_0^{-1} - b^{-1} - ik} \simeq \frac{1}{-b^{-1} - ik} . \tag{10.75}$$

10.7 Effective-range theory, resonance scattering

For centrally symmetric potentials the Fourier transforms in equation (10.71) can be expanded as

$$\begin{gathered} U(\mathbf{k}) = \int d\mathbf{r}\cdot U(\mathbf{r})e^{-i\mathbf{k}\mathbf{r}} = \\ = U_0 - (k^2/6)\int d\mathbf{r}\cdot U(r)r^2 + ... = U_0 - Ak^2 + ... , \end{gathered} \tag{10.76}$$

[9] A short-range potential has a weak k-dependence for $kb \ll 1$ and a decreasing oscillatory tail for $kb \gg 1$. The s-wave amplitude given by equation (10.71) is very close to the exact result for a spherical potential well.

where $U_0 = U(\mathbf{k}=0)$ and $A = (1/6)\int d\mathbf{r}U(r)r^2 > 0$. Such an expansion holds for $kb \ll 1$. The scattering amplitude given by equation (10.71) can then be written as $f = 1/(g - ik)$, where

$$g = -\frac{b+U_0}{bU_0} - 2A\frac{b+U_0}{bU_0^2}k^2 + 2A\left(\frac{b+U_0}{bU_0}\right)^2 k^2\cos\theta \ . \qquad (10.77)$$

Particularly interesting is the case $U_0 \sim -b$ in equation (10.77), when g vanishes and the scattering amplitude is very large; this corresponds to $V \simeq -\hbar^2/mb^2$. This is a resonance scattering (at low energy, $kb \ll 1$).[10] If there exists a shallow bound state, the scattering on this state is equivalent with the scattering on a potential equal to the energy of the state; it is a resonant scattering, where the particle spends much time on the bound state. We note that the contribution of the p-wave is of second order in comparison with that of the s-wave, so we can limit ourselves to the s-wave amplitude in equation (10.77). It can be written as

$$f_0 = \frac{1}{-a^{-1} + r_0k^2/2 - ik} \ , \qquad (10.78)$$

where

$$a = \frac{bU_0}{b+U_0} \qquad (10.79)$$

is the scattering length[11] and $r_0 = -4A(b+U_0)/bU_0^2$ is the parameter of the "effective range".[12] The s-wave amplitude is the effective range. For U_0 slightly below $-b$ ($U_0 < -b$) the scattering length is positive ($a > 0$) and r_0 is positive ($r_0 > 0$). The dependence of the total cross-section on energy can be followed experimentally, and the parameters a and r_0 can be determined, characterizing thus the potential. It is likely to exist, in this case, a bound state of small energy $\varepsilon = -\hbar^2\kappa^2/2m < 0$ and radius $R = 1/\kappa$. The real part of equation (10.78)

[10] E. Wigner, "Uber die Streuung von Neutronen an Protonen", Z. Phys. **83** 253 (1933).

[11] E. Fermi and L. Marshall, "Interference phenomena of slow neutrons", Phys. Rev. **71** 666 (1947).

[12] J. Schwinger, "Neutron scattering in ortho- and parahydrogen and the range of nuclear forces", Phys. Rev. **58** 1004 (1940); H. Bethe, "Theory of the effective range in nuclear scattering", Phys. Rev. **76** 38 (1949).

reads then $f_0^{-1} = -a^{-1} - r_0/2R^2 = -1/R$. This may provide another check for the two parameters a and r_0. Historically, this was the procedure employed for determining the scattering length $a \sim 4fm$ for the triplet neutron-proton scattering, corresponding to a bound state $-\varepsilon \sim 2MeV$ and a potential depth $V \sim -10MeV$ for a range $b \sim 1-2fm$ of the nuclear forces ($1fm = 10^{-15}m$).

Leaving aside the k^2-term in equation (10.78), and writing $-a^{-1} = -1/R = -\kappa$, the scattering amplitude reads

$$f_0 = -\frac{1}{\kappa + ik} = -(2m/\hbar^2)^{1/2}\frac{1}{\sqrt{|\varepsilon|} + i\sqrt{E}} \ . \tag{10.80}$$

This is Wigner's formula for resonance scattering on an upper discrete level.

For U_0 slightly above $-b$ ($-b < U_0 < 0$), the scattering length is negative ($a < 0$) and r_0 is also negative ($r_0 < 0$). There is not likely to exist now a bound state, so, the k^2-term in equation (10.78) can be left aside. Equation (10.80) reads now

$$f_0 = -\frac{1}{\kappa + ik} = -(2m/\hbar^2)^{1/2}\frac{1}{-\sqrt{\varepsilon} + i\sqrt{E}} \ , \tag{10.81}$$

where $\kappa = a^{-1} < 0$; $\varepsilon > 0$ is said to be a virtual energy level. Knowing the triplet scattering length and measuring the coherent scattering of neutrons from hydrogen molecule, the scattering length $a \leq -2fm$ was obtained for the singlet neutron-proton scattering.[13] This suggested the spin dependence of the nuclear forces, and the fact that the singlet neutron-proton nearly misses a bound state. It has a virtual energy level.

For a partial-wave $f_l = 1/(g_l - ik)$, the function g_l starts with $k^{-2l} = E^{-l}$, and continues with $k^{-2l+2} = E^{-l+1}$. Keeping the first two terms in such an expansion, the partial scattering amplitude reads

$$f_l = \frac{1}{cE^{-l}(-\varepsilon + E) - ik} \ , \tag{10.82}$$

where c and ε are two parameters. For $\varepsilon < 0$ we have a pole $E = -|\varepsilon|$, *i.e.* a discrete level. The scattering amplitude is, however, low,

[13]Similar negative scattering lengths are known for (singlet) neutron-neutron and proton-proton.

because of the small factor E^l. The resonance scattering of higher-order partial waves on bound states is low. For $\varepsilon > 0$, there is no bound state, but the resonance is high on the virtual level $\varepsilon > 0$. The scattering amplitude looks practically like $f = 1/(-ik)$ which, for low energies, means a very long lifetime. By equation (10.82), the width of this resonance is given by $\Delta E/\varepsilon \sim E^l k/k^2 = k^{2l-1}$. For low energy, this is a very sharp resonance. The centrifugal barrier leaves little chance for the motion to escape too soon. This is different from the resonance in s-wave.

10.8 Breit-Wigner formula

During collision, particles can be absorbed for a while on the target, and additional particles can be released by the target after a while, by disintegration. Additional contribution is then present in the scattering amplitude.

Equation (10.81) can be written generically as $f = (\hbar^2/2m)^{1/2}/\sqrt{E}$ (negative scattering length). Its variation is therefore $\delta f = -(\delta E/2k)/E$. Near a resonance level E_0 ($E_0 > 0$) the denominator E means $E - E_0$. On the other hand, the level E_0 is only a "quasi-discrete" level, *i.e.* it is represented as $E_0 - i\Gamma/2$, such that its uncertainty (the width) is $\delta E = \Gamma(\Gamma > 0)$. We get

$$\delta f = -\frac{\Gamma/2k}{E - E_0 + i\Gamma/2} \,. \tag{10.83}$$

This is the Breit-Wigner formula for resonance scattering on a "quasi-discrete" level.[14] Obviously, the width Γ must be much smaller than the levels separation. On the other hand, E is close to E_0.

The wavefunction of the "quasi-discrete" (and quasi-stationary) level goes like $\psi \sim \exp(-iE_0t/\hbar - \Gamma t/2\hbar)$, *i.e.* $|\psi|^2 \sim \exp(-\Gamma t/\hbar)$; the level decays with the probability $w = \Gamma/\hbar$ per unit time, and has a lifetime $\tau = \hbar/\Gamma = 1/w$. The decaying levels are positive, and their spectrum is, in fact, quasi-continuous. The second-order perturbation theory

[14] G. Breit and E. P. Wigner, "Capture of slow neutrons", Phys. Rev. **49** 519 (1936).

gives the energy correction

$$\delta E = \int \frac{|V_{n\nu}|^2 \, d\nu}{E_n - E_\nu + i0} \tag{10.84}$$

with usual notations, and, since $1/(x+i0) = P(1/x) - i\pi\delta(x)$, we get

$$\Gamma = 2\pi \int |V_{n\nu}|^2 \, \delta(E_n - E_\nu) d\nu \; . \tag{10.85}$$

The scattering amplitude reads then $f = f_0 + \delta f$, where f_0 is the potential part of the scattering amplitude derived before; δf is the resonance part.

For resonance level E_0 close to zero, its width must vanish, since $E \to 0$. It follows that in

$$\delta f = (\hbar^2/2m)^{1/2} \delta(1/\sqrt{E}) = -\frac{\hbar}{\sqrt{2m}} \frac{\delta E / 2\sqrt{E}}{E} \tag{10.86}$$

the width $\Gamma = \delta E$ must have the form $\Gamma = 2\gamma\sqrt{E}$, where γ is finite. The Breit-Wigner formula becomes

$$\delta f = -\frac{\hbar}{\sqrt{2m}} \frac{\gamma}{E - E_0 + i\gamma\sqrt{E}} \; . \tag{10.87}$$

For $E \to 0$ this δf is now finite, in contrast with equation (10.83).

If $E \ll \gamma^2$ it is easy to see that in equation (10.87) we may neglect the E-term. Equation (10.87) recovers then the resonance scattering on a vanishing level, either positive or negative. Indeed, the scattering amplitude can be writen as

$$\delta f = -\frac{1}{ik - \sqrt{2m}E_0/\hbar\gamma} \; . \tag{10.88}$$

We define $-\kappa = \sqrt{2m}E_0/\hbar\gamma$, and get $\delta f = -1/(\kappa + ik)$. This corresponds to a resonance scattering on a virtual level for $\kappa < 0$ ($a < 0$, according to equation (10.81)), or a true discrete level for $\kappa > 0$ (according to equation (10.80).

10.9 Bound states

Let us seek bound states of negative energy $E = -\hbar^2k^2/2m$ of a particle of mass m in a short-range potential well $-|V|$, where k is the "wavevector" associated with energy E.[15] Schrodinger's equation reads

$$(\Delta - k^2)\psi = -4\pi U\psi \ , \tag{10.89}$$

where $U = m\,|V|\,/2\pi\hbar^2$. Solutions of this equation can be represented by

$$\psi(\mathbf{r}) = \int d\mathbf{r}' \cdot U(\mathbf{r}')\psi(\mathbf{r}')\frac{e^{-k|\mathbf{r}-\mathbf{r}'|}}{|\mathbf{r}-\mathbf{r}'|} \ , \tag{10.90}$$

which is the integral Schrodinger equation. The asymptotic behaviour of this solution reads

$$\psi(\mathbf{r}) = \int d\mathbf{r}' \cdot U(\mathbf{r}')\psi(\mathbf{r}')e^{\mathbf{k}'\mathbf{r}'} \cdot \frac{e^{-k\mathbf{r}}}{r} \ , \tag{10.91}$$

where $\mathbf{k}' = k\mathbf{r}/r$.

For small values of the potential U equation (10.90) can be approximated by

$$\psi(\mathbf{r}) \simeq \int d\mathbf{r}' \cdot U(\mathbf{r}')\psi(\mathbf{r}') \cdot \frac{e^{-kb}}{b} \ , \tag{10.92}$$

where b is the range of the potential. We get immediately the energy level given by

$$k^2 = \frac{1}{b^2}\ln^2[\frac{1}{b}\int d\mathbf{r}'U(\mathbf{r}')] \ . \tag{10.93}$$

We can see that it does not exist unless the potential U exceeds the minimum value given by

$$1 = \frac{1}{b}\int d\mathbf{r}'U_0(\mathbf{r}') \simeq 4\pi b^2U_0/3 \ , \tag{10.94}$$

where U_0 is the mean value of the potential. For this threshold value of the potential the energy level is vanishing. For an increase δU the

[15] M. Apostol, "Bound states in short-range potentials", J. Theor. Phys. **130** (2006).

energy level increases (in absolute value) by

$$k^2 \simeq \frac{1}{b^2}(\delta U/U_0)^2 = (4\pi/3)U_0(\delta U/U_0)^2 \ . \tag{10.95}$$

For deeper potential wells we notice that the main contribution to the eigenvalues equation (10.90) comes from $\mathbf{r}' \sim \mathbf{r}$. Consequently we write it as

$$\psi(\mathbf{r}) = U \int d\boldsymbol{\rho} \cdot \psi(\mathbf{r} + \boldsymbol{\rho}) \frac{e^{-k\rho}}{\rho} \ , \tag{10.96}$$

where U is the mean potential and integration is now performed over a range β which depends on $\mathbf{r}$. Equation (10.96) admits plane waves $e^{i\kappa\mathbf{r}}$ as approximate solutions. Carrying out the integration we get

$$1 = 4\pi U \left\{ \frac{1}{\kappa^2 + k^2} - \frac{\cos\kappa\beta + (k/\kappa)\sin\kappa\beta}{\kappa^2 + k^2} e^{-k\beta} \right\} \ , \tag{10.97}$$

for $\kappa\beta \gg 1$. At the same time, equation (10.90) gives for $\mathbf{r} = 0$

$$1 = 4\pi U \left\{ \frac{1}{\kappa^2 + k^2} - \frac{\cos\kappa b + (k/\kappa)\sin\kappa b}{\kappa^2 + k^2} e^{-kb} \right\} \ . \tag{10.98}$$

From equations (10.97) and (10.98), for infinitely large potential wells $\beta, b \to \infty$ we get the free, continuous spectrum $k^2 = 4\pi U - \kappa^2$, as expected.

Equation (10.97) should not depend on $\mathbf{r}$ (through β). For this, its derivative with respect to β must vanish. It gives $\kappa\beta = n\pi$ for all non-vanishing integers $n \neq 0$. The only way to satisfy both equations (10.97) and (10.98) is to put $\beta = b$ and quantize κ through $\kappa b = n\pi$. We get the dispersion relation

$$1 = 4\pi U \left\{ \frac{1}{\kappa^2 + k^2} - \frac{(-1)^n}{\kappa^2 + k^2} e^{-kb} \right\} \ . \tag{10.99}$$

Making use of equation (10.94), it is easy to see that this equation has solutions only for odd integers n. We write therefore the dispersion relations (10.99) as

$$e^{-kb} + 1 = \frac{\kappa^2 + k^2}{4\pi U} \ , \tag{10.100}$$

for $\kappa b = (2n+1)\pi$, $n = 0, 1, 2, ...$, $4\pi Ub^2 \geq 3$ and $\kappa b \gg 1$. We note that κ may run up to a maximum value given by $(\kappa_m b)^2 = 8\pi Ub^2$.

For $Ub^2 \gg 1$ the upper bound states are dense; from equation (10.100), their energy is given by $kb = 2 - \kappa^2/4\pi U$. With $\kappa = \kappa_m - \delta$ and $\delta b = 2\pi\delta n$ we get the top spectrum $-E = (2\pi\hbar^2/mb^2)^2(\delta n)^2/|V|$ for a deep potential well.

For deep potential wells ($Ub^2 \gg 1$) the above spectrum is applicable also for the ground state (high values of k), correspondig to the lowest $n = 0$, *i.e.* $\kappa b = \pi$. The ground state energy is easily obtained from equation (10.100) in this case as $-E_0 = |V| - \pi^2\hbar^2/2mb^2$.

The spectrum give by equation (10.100) is also applicable for shallow U, when there is only one bound state. It is given by $kb = 2 - \kappa^2/4\pi U$ for $\kappa b = \pi$. The critical depth is given by $U_0 b^2 = \pi/8$, which differs slightly from equation (10.94) ($U_0 b^2 = 3/4\pi$). The level goes like $kb = 2(\delta U/U_0)$, or $k^2 = (32/\pi)U_0(\delta U/U_0)^2$, which again differs from equation (10.95) by a numerical factor.

According to the derivation given above, the approximate eigenstates are plane waves, with spatial degeneracy. They are more appropriate for s-waves for central potentials, as the l-waves depend more on the details of the potential shape. For deep potential wells the number of energy levels is $\sim (8\pi Ub^2)^{1/2}$ (so that the mean distance between energy levels is $\delta E \sim (Ub^2)^{1/2}$) and the number of states is $\sim 4\pi(8\pi Ub^2)^{3/2}/3$. They read $(|V|/\varepsilon)^{1/2}$ and $\sim (|V|/\varepsilon)^{3/2}$, respectively, where $\varepsilon = \hbar^2/mb^2$ is the localization energy in the potential well.

10.10 Fast particles

In the high-energy limit $k \to \infty$ the second term in equation (10.71) goes to zero ($U(\mathbf{k}) \to 0$). The first term in equation (10.71) (Born's amplitude $-U(\mathbf{q})$) vanishes also for any finite θ in $q = 2k \sin\theta/2$ in the limit $k \to \infty$. It follows that the scattering proceeds mainly forwards, $\theta \to 0$, such that $q \simeq k\theta$ remains finite and small for $k \to \infty$. The small momentum transfer $\mathbf{q}$ is perpendicular to $\mathbf{k}$. Born's amplitude reads then

$$U(\mathbf{q}) = \int d\mathbf{r} \cdot U(\mathbf{r}) e^{-i\mathbf{q}\mathbf{r}} = \int d\boldsymbol{\rho} \cdot [\int_{-\infty}^{+\infty} dz \cdot U(z, \boldsymbol{\rho})] e^{-i\mathbf{q}\boldsymbol{\rho}} \ , \quad (10.101)$$

where $\boldsymbol{\rho}$ is the transverse variable with respect to the wavevector $\mathbf{k}$. For centrally symmetric potentials, the integral in equation (10.101) can be estimated by noting that the oscillating exponential is not vanishing only over a radius ρ, given by $\langle (\mathbf{q}\rho)^2 \rangle = q^2\rho^2/2 = 2$, *i.e.* $\rho = 2/q = 2/k\Delta\theta$. The corresponding area is therefore $\pi\rho^2 = 4\pi/k^2(\Delta\theta)^2$. We get[16]

$$U(\mathbf{q}) = \frac{4\pi}{k^2(\Delta\theta)^2} \int_{-\infty}^{+\infty} dz \cdot U(z, 0) \ . \quad (10.102)$$

Similarly, the total cross-section is given by

$$\begin{aligned} \sigma &= \int |U(\mathbf{q})|^2 \, d\Omega = \\ &= \int dz dz' d\boldsymbol{\rho} d\boldsymbol{\rho} \cdot U(z, \boldsymbol{\rho}) U(z', \boldsymbol{\rho}') e^{-i\mathbf{q}(\boldsymbol{\rho}-\boldsymbol{\rho}')} d\Omega \ , \end{aligned} \quad (10.103)$$

or

$$\sigma = \frac{4\pi^2}{k^2} \int d\boldsymbol{\rho} \cdot (\int dz U)^2 \ , \quad (10.104)$$

and the optical theorem gives $Imf(0) = k\sigma/4\pi$. One can see that $Imf(0)$ is a second-order correction to Born's amplitude in equation (10.71). Equation (10.103) is the classical expression of the cross-section integrated with respect to the impact parameter ρ.

Obviously, the forward scattering is the eikonal regime, and the Schrodinger equation (10.61) can be written as

$$[\Delta + k^2 - 4\pi U(\mathbf{r})]\psi = (\Delta + \kappa^2)\psi = 0 \ , \quad (10.105)$$

where $\kappa^2 = k^2 - 4\pi U$. For large k the solution reads

$$\psi = \exp(i \int^z dz \cdot \kappa) \ , \quad (10.106)$$

[16] $\langle \exp(i\varphi) \rangle = 1 - \langle \varphi^2 \rangle /2 = 0$, *i.e.* $\langle \varphi^2 \rangle = 2$, which means $\langle (\mathbf{q}\rho)^2 \rangle = q^2\rho^2/2 = 2$, and $\rho = 2/q$; the area factor is therefore $\pi\rho^2 = 4\pi/q^2 = 4\pi/k^2(\Delta\theta)^2$.

providing the potential has a smooth variation (quasi-classical approximation). Indeed, the terms neglected in the laplacian of equation (10.105) are of the order of $\partial\kappa/\partial z$ which must be compared to κ^2 (and similar derivatives with respect to the other coordinates). This means $\partial\lambda/\partial z \ll 1$, where λ is the wavelength; this is the typical condition of validiy for the quasi-classical approximation. It can also be written as $\partial U/\partial z \ll \kappa^3$, or $|U| \ll (\kappa b)\kappa^2$, a condition which may be read as $|U| \ll (kb)k^2$, since k is much greater than κ. However, it is worth noting that this condition is much weaker that $|U|/k^2 \ll 1$, in virtue of $kb \gg 1$. Therefore, we may use equation (10.106) with this weaker condition, without resorting to approximating κ by $k - 2\pi U/k$. It is worth noting that such a first-order approximation in equation (10.106) renders in fact $\psi = \exp(i\int^z dz \cdot \kappa) = \exp(ikz)[1 - (2\pi i/k)\int^z dz \cdot U]$, and it would not give an imaginary part of the scattering amplitude.

Equation (10.106) can also be written as

$$\psi = e^{ikz}S(z,\rho) \ , \qquad (10.107)$$

where

$$S(z,\rho) = \exp\left[i\int^z dz \cdot (\kappa - k)\right] = $$
$$= \exp\left(-\frac{4\pi i}{k}\int^z dz \cdot \frac{U}{\sqrt{1-4\pi U/k^2}+1}\right) \ . \qquad (10.108)$$

One can see from equation (10.108) that the phase variation is of the order of $|U|\,b/k$. It holds for $|U| \ll k^2$. The quasi-classical approximation is still valid ($|U| \ll k^2(kb)$), even for $kb \sim 1$, providing this stronger condition is satisfied. Then, we get $|U|\,b/k \ll 1$, or, since $U = mV/2\pi\hbar^2$, $|V| \ll (\hbar^2/mb^2)kb = \hbar v/b$, where v is the particle velocity.[17] This condition tells that U can be treated as a perturbation, and, again, ψ can be approximated by $\exp(ikz)[1 - (2\pi i/k)\int^z dzU]$.

The scattering amplitude can be obtained from equation (10.62), written as

$$\varphi(\mathbf{r}) = -\int d\mathbf{r}' \cdot U(\mathbf{r}')\psi(\mathbf{r}')\frac{e^{ik|\mathbf{r}-\mathbf{r}'|}}{|\mathbf{r}-\mathbf{r}'|} \ , \qquad (10.109)$$

[17] For a Coulomb potential α/r this means $|\alpha| \ll \hbar v$, *i.e.* the Bohr radius $\hbar^2/m|\alpha|$ is much larger than the wavelength λ. The Coulomb potential may then be treated as a perturbation to the scattering in a short range potential.

which becomes asymptotically

$$\varphi(\mathbf{r}) = -\int d\mathbf{r}' \cdot U(\mathbf{r}')e^{-i\mathbf{q}\mathbf{r}'}S(z',\boldsymbol{\rho}') \cdot \frac{e^{ikr}}{r} \ . \tag{10.110}$$

Hence, the scattering amplitude

$$f = -\int d\rho dz \cdot U(z,\boldsymbol{\rho})S(z,\boldsymbol{\rho})e^{-i\mathbf{q}\rho} \ , \tag{10.111}$$

where $S(z,\boldsymbol{\rho})$ is given by equation (10.108) with the lower limit of integration $-\infty$. One can see that the first term in equation (10.111) is Born's amplitude. We can define

$$\int dz \cdot U(z,\boldsymbol{\rho})S(z,\boldsymbol{\rho}) = (ik/2\pi)[S(\boldsymbol{\rho}) - 1] \tag{10.112}$$

and write the scattering amplitude as

$$f = -(ik/2\pi)\int d\boldsymbol{\rho} \cdot [S(\boldsymbol{\rho}) - 1]e^{-i\mathbf{q}\rho} \ . \tag{10.113}$$

Then we get $Imf(0) = -(k/2\pi)\int d\rho \cdot Re[S(\boldsymbol{\rho}) - 1]$ and the total cross section

$$\sigma = -2\int d\boldsymbol{\rho} \cdot Re[S(\boldsymbol{\rho}) - 1] \ . \tag{10.114}$$

As regards an advanced computation of $S(\boldsymbol{\rho})$ in equation (10.112) we note that it is $S(+\infty,\boldsymbol{\rho})$ providing $S(z,\boldsymbol{\rho})$ in equation (10.108) is approximated by

$$S(z,\boldsymbol{\rho}) = \exp\left(-\frac{2\pi i}{k}\int_{-\infty}^{z} dz \cdot U\right) \ ; \tag{10.115}$$

indeed, we have then, formally, $US = (ik/2\pi)dS/dz$ and the integration in equation (10.112) gives $(ik/2\pi)[S(+\infty,\boldsymbol{\rho}) - 1]$. However, as it was said above, retaining only U in equation (10.108) amounts to

$$S(\boldsymbol{\rho}) = S(+\infty,\boldsymbol{\rho}) = 1 - \frac{2\pi i}{k}\int_{-\infty}^{+\infty} dz \cdot U \ , \tag{10.116}$$

which leads to $Re(S - 1) = 0$, *i.e.* no imaginary part in the scattering amplitude, and no scattering according to the optical theorem.

The actual computation in equation (10.111) gives in fact

$$f = \frac{1}{-k^2(\Delta\theta)^2/4\pi A - ik(\Delta\theta)^2/4} , \tag{10.117}$$

where $A = \int dz \cdot U(z,0)$ and $S(z,0) = 1 - (2\pi i/k)\int^z dz \cdot U(z,0)$ is used. The partial-wave amplitudes f_l in the expansion $f = \sum(2l + 1)f_l P_l(\cos\theta)$, where P_l are the Legendre polynomials, are given by

$$f_l = \frac{1}{-k^2/\pi A - ik} , \tag{10.118}$$

in accordance with the optical theorem.

Similarly, making use of equation (10.111), we get

$$f(0) = -\int d\rho \cdot A(1 - \frac{i\pi}{k}A) , \tag{10.119}$$

where $A = \int dz \cdot U(z,\boldsymbol{\rho})$ and $Imf(0) = (\pi/k)\int d\boldsymbol{\rho} \cdot A^2$. Also, by equation (10.111), the total cross-section reads

$$\begin{aligned} \sigma = \pi(\Delta\theta)^2 \int d\boldsymbol{\rho} d\boldsymbol{\rho}' dz dz' \cdot USU'S' e^{-i\mathbf{q}(\boldsymbol{\rho}-\boldsymbol{\rho}')} = \\ = \tfrac{4\pi^2}{k^2} \int d\boldsymbol{\rho} (\int dz \cdot US)^2 , \end{aligned} \tag{10.120}$$

which is $4\pi/k)Imf(0)$ in agreement with the optical theorem and Born's scattering amplitude.

It is worth estimating the contribution of the second term in equation (10.71) for high-energy scattering. We employ the following successive approximations

$$\begin{aligned} U(\mathbf{k}) = \int d\mathbf{r} \cdot U(\mathbf{r}) e^{-i\mathbf{k}\mathbf{r}} = \int d\boldsymbol{\rho} dz \cdot U(z,\boldsymbol{\rho}) e^{-ikz} = \\ = \pi b^2 \int dz \cdot U(z,0) e^{-ikz} = (\pi b^2/bk) \int dz \cdot U(z,0) \end{aligned} \tag{10.121}$$

in this case, and get

$$\begin{aligned} U(\mathbf{k}') = \int d\boldsymbol{\rho} dz \cdot U(z,\boldsymbol{\rho}) e^{-ikz} e^{-i\mathbf{q}\boldsymbol{\rho}} = \\ = [4\pi/k^2(\Delta\theta)^2](1/bk) \int dz \cdot U(z,0) . \end{aligned} \tag{10.122}$$

Then, $U(\mathbf{k}=0)$ can be neglected in the denominator of equation (10.71) and $b'^{-1}=b^{-1}+ik$. The contribution of b^{-1} can also be neglected within this approximation, so we get finally

$$f=-U(\mathbf{q})+ik\frac{4\pi^2}{k^4(\Delta\theta)^2}[\int dz\cdot U(z,0)]^2 \ , \tag{10.123}$$

which coincides with equation (10.117). We may say that the representation given by equation (10.71) for the scattering amplitude can be viewed as a fairly adequate one, within the given conditions; in the short-wavelengths limit, equation (10.71) gives the same results as the quasi-classical theory of scattering, optical theorem included, for the second-order of the perturbation theory. In the intermediate range $kb\sim 1$ the main contribution to equation (10.71) comes from Born's amplitude. Its partial waves must then be corrected by $f_l\rightarrow f_l/(1-ikf_l)$ in order to satisfy the optical theorem. This is a small correction, of the order of the second term in equation (10.71).

10.11 Coulomb scattering

Schrodinger's equation for Coulomb potential $V=\alpha/r$ reads

$$(\Delta-\frac{2}{ar})\psi=-k^2\psi \ , \tag{10.124}$$

where $a=\hbar^2/m\alpha$ is the Coulomb localization length (Bohr radius) and $k^2=2mE/\hbar^2>0$ is the wavevector of the energy E for the scattering of a particle with mass m (we assume first $\alpha>0$). According to equation (10.124), the localization energy $\hbar^2/ma^2$ competes with the Coulomb energy α/a; in order to calibrate the results we may view $1/ka$ as a perturbation parameter, where $ka=\hbar v/\alpha\gg 1$, where v is the velocity of the particle; it means $\alpha\ll\hbar^2k/m=\hbar v$.

Writing $r=\sqrt{\rho^2+z^2}$, one can see that there is a singularity for $\rho\rightarrow 0$. This happens either for $z=r$ or for $z=-r$. We choose $z=r$, corresponding to an incoming plane wave e^{ikz} for $z\rightarrow-\infty$. Equation (10.124) becomes asymptotically

$$d^2\psi_i/dz^2-(2/ar)\psi_i=-k^2\psi_i \ , \tag{10.125}$$

whose solution is

$$\psi_i = e^{ikz+(i/ka)\ln k(r-z)} \; . \tag{10.126}$$

One can see that the plane wave is distorted by forward scattering because of the long range of the Coulomb potential. In addition, the distortion is vanishing for $ka \gg 1$ for any finite scattering angle. A series expansion in powers of $1/a$ corrects not only the phase of the plane wave, but also its amplitude.

Leaving aside the centrifugal potential, the spherical wave $\psi_s = \chi/r$ satisfies

$$\chi'' - (2/ar)\chi = -k^2\chi \; ; \tag{10.127}$$

in compliance with the scattering boundary condition, we write formally $r \rightarrow A - z$, where $A = const \rightarrow r$. The wavefunction becomes $\psi_s = \chi/(A-z)$ and equation (10.127) reads now

$$d^2\chi/dz^2 - (2/ar)\chi = -k^2\chi \; ; \tag{10.128}$$

its solution is

$$\chi = e^{ik(A-z)-(i/ka)\ln k(r-z)} \rightarrow e^{ikr-(i/ka)\ln k(r-z)} \tag{10.129}$$

and

$$\psi_s = \frac{1}{A-z}e^{ikr-(i/ka)\ln k(r-z)} \rightarrow \frac{1}{r-z}e^{ikr-(i/ka)\ln k(r-z)} \; . \tag{10.130}$$

The full scattering solution $\psi = \psi_i + \psi_s$ reads

$$\psi = e^{ikz+(i/ka)\ln k(r-z)} + f \cdot e^{ikr-(i/ka)\ln(2kr)}/r \; , \tag{10.131}$$

where the scattering amplitude can be written as

$$f = -\frac{1}{2k^2 a \sin^2\theta/2}e^{-(2i/ka)\ln(\sin\theta/2)} \; ; \tag{10.132}$$

the prefactor $1/ka$ in equation (10.132) is chosen by dimensional arguments, such as to agree with a $1/a$-expansion. Additional phase factors depending on k may also be present in the scattering amplitude.[18]

[18] M. Apostol, "Coulomb scattering", J. Theor. Phys. **131** (2006).

Equation (10.132) gives the Rutherford cross-section $d\sigma = 1/4k^4a^2 \cdot \sin^4\theta/2$; it does not depend on the Born approximation or the partial-wave expansion. We can see that phase shifts of the type δ_l do not occur in the Coulomb wavefunction and there exists a singularity of the scattering amplitude for $\theta = 0$. If $\alpha < 0$ (attractive potential), the relevant phase factors change the sign and the scattering amplitude becomes the complex conjugate of the expression given by equation (10.132).

If an additional (short-range) potential is present, as the nuclear potential for instance, then it contributes its own scattering amplitude. For the s-wave at resonance, it reads $f_0 = a_0/(1 - ia_0k)$, where a_0 is the (large) scattering length (and $k \sim 0$). This scattering amplitude can also be written as $f_0 = a_0e^{i\delta}$ for $a_0k \sim \delta$, such that an interference term appears in the scattering cross-section, of the form $\cos\left(\frac{2}{ka}\ln\sin\theta/2 + \delta\right)$. This interference term may serve to correct the additional scattering length a_0 (through δ) by terms of the form $(1/a)\ln a$, for right-angle scattering where $a\sin\theta/2 \sim const.$

10.12 Electron-atom scattering

If the incident electrons are faster than the electrons inside the atom we may view the atom as "inert" with respect to the electron scattering, and the scattering amplitude as being averaged over the electron motion inside the atom. The potential φ is given by

$$\Delta\varphi = 4\pi Ze\delta(\mathbf{r}) - 4\pi en \quad , \tag{10.133}$$

where Z is the atomic number of the atom, n is the electron density inside the atom and e is the electron charge; we need the Fourier transform $\varphi(\mathbf{q})$ of the potential φ for the Born scattering cross-section, where $\mathbf{q}$ is the momentum transfer, $q = 2k\sin\theta/2$, $k = \sqrt{2mE/\hbar^2}$, E being the energy of the incident electrons and θ being the scattering angle (m denotes the mass of the electron); from equation (10.133) we get $\varphi(\mathbf{q}) = -4\pi e[Z - n(\mathbf{q})]/q^2$, where $n(\mathbf{q})$ is the Fourier transform of the electron density in the atom; it is denoted by $F(\mathbf{q}) = n(\mathbf{q}) = \int d\mathbf{r}n(\mathbf{r})e^{-i\mathbf{q}\mathbf{r}}$ and is called the atomic form factor. We get the cross-section

$$d\sigma = \frac{4m^2e^4}{\hbar^4 q^4}[Z - F(\mathbf{q})]^2 d\Omega = 4a_H^2[Z - F(\mathbf{q})]^2 d\Omega \ , \qquad (10.134)$$

where a_H is the Bohr radius. For heavy atoms the atomic form factor can be computed by using the Thomas-Fermi theory.

For small angles we may expand $F(\mathbf{q})$ in powers of q, $Z - F(\mathbf{q}) \simeq \frac{1}{6}q^2 \int d\mathbf{r} n r^2$; the cross-section is independent of the scattering angle and is given by the mean square distance of the atomic electrons from the nucleus. For large q the form factor is vanishing and we get the Coulomb scattering from the nucleus.

It is easy to see that $(1 - \cos\theta)d\sigma$ is related to the electrons scattered outside the incident direction; this quantity is called the transport cross-section $d\sigma_{tr}$; since $qa_H \simeq (v/v_0)\theta$, where v is the velocity of the incident electrons and v_0 is the velocity of the atomic electrons, we have

$$\sigma_{tr} \simeq 4\pi(Ze^2/mv^2)^2 \ln \gamma\hbar v/e^2) \qquad (10.135)$$

for $v_0/v \ll \theta \ll 1$, where γ is an undetermined constant. This is related to Bethe's stopping power of charged particles in matter.[19]

10.13 Collision between like particles

If the colliding particles are identical, the scattering wavefunction at large distances must be symmetrized, according to their total spin;[20] it reads

$$\psi = e^{ikz} \pm e^{-ikz} + [f(\theta) \pm f(\pi - \theta)] \frac{e^{ikr}}{r} \ ; \qquad (10.136)$$

the cross-section is

$$d\sigma = |f(\theta) \pm f(\pi - \theta)|^2 d\Omega \ ; \qquad (10.137)$$

[19] H. Bethe, "Zur Theorie des Durchgangs schneller Korpuskularstrahlen durch Materie", Ann. Phys. **5** 325 (1930).

[20] N. F. Mott, "The collision between two electrons", Proc. Roy. Soc. **A126** 259 (1930).

it includes interference terms, due to the exchange of the particles; the colliding particles are indiscernable. If the spin is not defined, we must average over spin states. Of the total number of $(2s+1)^2$ spin states of two identical particles with spin s, $s(2s+1)$ states correspond to an even total spin and $(s+1)(2s+1)$ states correspond to an odd total spin if s is half-integral, and viceversa for an integral s; for instance, we have

$$d\sigma = \frac{s}{2s+1}d\sigma_s + \frac{s+1}{2s+1}d\sigma_a \tag{10.138}$$

for a half-integral total spin, where the cross-sections $d\sigma_{s,a}$ correspond to symmetrized/antisymmetrized scattering amplitudes. For two electrons (interacting by Coulomb's law e^2/r) with total spin 1/2, making use of equation (10.132), the cross-section is

$$\begin{gathered} d\sigma = (e^2/m_0v^2)^2\{\tfrac{1}{\sin^4\frac{\theta}{2}} + \tfrac{1}{\cos^4\frac{\theta}{2}} - \\ \tfrac{1}{\sin^2\frac{\theta}{2}\cos^2\frac{\theta}{2}}\cos\left(\tfrac{e^2}{\hbar v}\ln\tan^2\frac{\theta}{2}\right)\}d\Omega \quad , \end{gathered} \tag{10.139}$$

where m_0 is the mass and v is the velocity of the electron. These formulae are valid for the center of mass at rest; if one particle is at rest we must replace θ by 2θ.

10.14 Scattering in two dimensions

The scattering wavefunction in two dimensions has the asymptotic form

$$\psi = e^{ikz} + f(\varphi)\frac{e^{ikr}}{\sqrt{-ir}} \tag{10.140}$$

(where $-i$ is introduced for convenience); the cross-section is $d\sigma = \mid f \mid^2 d\varphi$. For centrally symmetric potentials the free wavefunctions are the Bessel functions (up to factors $e^{im\varphi}$); in the presence of short-range potentials their asymptotic behaviour is

$$Q_m(r) \simeq i^m\sqrt{\frac{2}{\pi kr}}\sin\left[kr - (m-1/2)\pi/2 + \delta_m\right] \quad , \tag{10.141}$$

where δ_m are phase shifts; we get

$$\psi = \sum_{m=-\infty}^{+\infty} e^{i\delta_m} Q_m(r) e^{im\varphi} \tag{10.142}$$

and the scattering amplitude

$$f(\varphi) = \frac{1}{i\sqrt{2\pi k}} \sum_m \left(e^{2i\delta_m} - 1\right) e^{im\varphi} \ ; \tag{10.143}$$

the total cross-section is

$$\sigma = \sum_m \sigma_m \ , \ \ \sigma_m = \sigma_{-m} = (4/k)\sin^2\delta_m \ ; \tag{10.144}$$

the optical theorem reads $Imf(0) = \sqrt{k/8\pi}\sigma$.
For scattering at small angles we get

$$f = \frac{1}{i}\sqrt{\frac{k}{2\pi}} \int dx \left[S(x) - 1\right] e^{-iqx} \ \ , \tag{10.145}$$

by analogy with equation (10.113).
The Aharonov-Bohm effect is an application of this scattering theory in two dimensions.[21] Let us assume that a uniform magnetic field H is enclosed in a cylindrical region of radius a and infinite length along the y-direction; from the flux $\Phi = \int H dxdz$ we get the potential vector $A = (\Phi/2\pi) grad\varphi$, where φ is the polar angle in the (x, z)-plane; the wavefunction of a charge e moving along the z-direction is

$$\psi = e^{ikz} e^{\frac{ie\Phi\varphi}{2\pi\hbar c}} \ . \tag{10.146}$$

Since this wavefunction is multi-valued, a cut is necessary along the semi-axis $z > 0$; for small momentum transfer $q \simeq k\varphi$, $qa \ll 1$, $\varphi \ll 1$, and for large distances $x \simeq 1/q \gg a$ and $z \gg \mid x \mid$ the wavefunction can be written as $\psi = e^{ikz} F$, where

$$F = \begin{cases} e^{-\frac{ie\Phi}{2\hbar c}} \ , & x > 0 \ , \\ e^{\frac{ie\Phi}{2\hbar c}} \ , & x < 0 \ . \end{cases} \tag{10.147}$$

[21]Y. Aharonov and D. Bohm, "Significance of the electromagnetic potentials in the quantum theory", Phys. Rev. **115** 485 (1959).

Making use of equation (10.145), we have

$$f=\frac{1}{i}\sqrt{\frac{k}{2\pi}}\left[e^{-\frac{ie\Phi}{2\hbar c}}\int_0^{\infty}e^{-iqx}dx+c.c.\right] , \tag{10.148}$$

or

$$f=\frac{i}{q}\sqrt{\frac{2k}{\pi}}\sin\frac{e\Phi}{2\hbar c} \tag{10.149}$$

(by using a cutoff). The scattering cross-section is

$$d\sigma=\mid f\mid^2 d\varphi=\frac{2}{\pi k}\sin^2\frac{e\Phi}{2\hbar c}\cdot\frac{d\varphi}{\varphi^2} . \tag{10.150}$$

We can see that the electronic wavefunctions separated by a confined magnetic field interfere, due to non-zero vector potential outside the region of the magnetic field.

10.15 Concluding remarks

Some general properties of the scattering amplitude can be derived by viewing the energy E as a complex variable.[22] The scattering amplitude is analytic everywhere, except for a discrete set of poles E_n on the negative-energy semi-axis and a cut along the positive-energy semi-axis. For large energies the Born amplitude f_B does not include anymore the angle factor, so it is related to the forward scattering amplitude $f(0,E)$ by the dispersion relation

$$f(0,E)=f_B+\frac{1}{\pi}\int_0^{\infty}dE^{'}\frac{Imf(0,E^{'})}{E^{'}-E}+\sum_n\frac{d_n}{E-E_n} , \tag{10.151}$$

where d_n are the residues of the poles E_n (and E is $E+i0$). The Fourier transform $\chi(\mathbf{k}^{'})$ of the scattered wave $f(\mathbf{k},k\mathbf{r}/r)e^{i\mathbf{kr}}/r$ is

$$\chi(\mathbf{k}^{'})=4\pi\frac{f(\mathbf{k},\mathbf{k}^{'})}{k^{'2}-k^2-i0} . \tag{10.152}$$

[22] L. Landau and E. Lifshitz, *Course of Theoretical Physics*, vol. 3, *Quantum Mechanics (Non-relativistic Theory)*, Elsevier (1977).

If we regard the scattering amplitude as a function of complex angular momentum l (for real E), then its poles for integral l are related to bound states and resonance scattering; these are called Regge poles.[23]

The scattering may depend on spin, as in case of nuclear forces; then, a spin-orbit factor **ls** is included in the scattering amplitude; the action of **l** on the Legendre polynomials can be computed by using the orbital momentum in spherical coordinates; the scattering amplitude may depend in this case on the azimuthal angle too.

The scattering may take place with a change in the internal state of the colliding particles; then, the scattering is inelastic, and we are in the presence of reactions, with reaction channels; there exists an exchange of energy, and the scattering wavefunction is not stationary anymore; the phase shifts δ_l in $S_l = e^{2i\delta_l}$ may acquire an undetermined imaginary part; the cross-section implies now $1- \mid S_l \mid^2$ instead of $\mid 1 - S_l \mid^2$. The resonance scattering is related to inelastic scattering, and Breit-Wigner scattering formulae are applicable; as, for instance, in slow neutron collisions, where a compound nucleus is formed.[24] An interesting inelastic collision is the ionization of an atom by electrons with energy slightly greater than the ionization energy.[25] The energy loss of electrons passing through matter is another example,[26] or the scattering of neutrons by nuclei.[27]

[23] T. Regge, "Introduction to complex orbital momenta", Nuovo Cim. **14** 951 (1959).

[24] N. Bohr, "Neutron capture and nuclear constitution", Nature **137** 344 (1936).

[25] G. H. Wannier, "The threshold law for single ionization of atoms or ions by electrons", Phys. Rev. **90** 817 (1953).

[26] H. Bethe, "Zur Theorie des Durchgangs schneller Korpuskularstrahlen durch Materie", Ann. Phys. **5** 325 (1930).

[27] E. Fermi, "On the recombination of neutrons and protons", Phys. Rev. **48** 570 (1936); J. Schwinger and E. Teller, "The scattering of neutrons by ortho- and parahydrogen", Phys. Rev. **51** 775 (1937); **52** 286 (1937).

11 Measurement Process

11.1 Introduction

Initially, in Newton's classical Mechanics, the measurements were made directly, by comparing, with naked eye, lengths, weigths (masses) or times with conventional units like meter, kilogram or second. Later, in Thermodynamics and Statistical Physics, the same procedure was used with temperature degrees; the law of the ideal gases provided an absolute temperature scale and the Kelvin degree. Later on, with the advent of the Electromagnetism and the atomistic conception of matter, the measurement philosophy changed, since the physical objects become too many and the physical quantities too small to be accessible by a direct comparison with units. First, by Dalton's law of chemical reactions in simple and multiple proportions, it was realized that the hydrogen atom is the lightest, and all the other atoms are multiple of the hydrogen atom; a quantity of substance with mass numerically equal in grams with the hydrogen multiple of that substance's atom is the mol; it contains Avogadro's number $\simeq 6 \times 10^{23}$ atoms (or molecules). Faraday's law of electrolysis and Thomson's gas discharges pointed out the existence of the electron. The electric charge of the electron was measured by Millikan, by compensating Coulomb force with mechanical forces; it is $e \simeq -4.8 \times 10^{-10} esu$, so, the esu unit was introduced for the electric charge. Knowing the charge, we can know the electric field and the electric potential (*e.g.*, for a capacitor); as well as the electric current and the magnetic field produced by an electric current. Knowing the fields we can measure the ratio e/m for the electron, whence the electron mass $m \simeq 10^{-27} g$. In all these inferences we need the equations of the Electromagnetism. The direct comparison with units disappeared, and the measuring process becomes indirect. Similarly, the light wavelengths were measured by interference, the light polarization was determined by means of crystals, and the light energy (flux) was determined by means of a black body. We moved far away from the direct comparison with naked eye. By indirect methods, using equations, the constant of Newton's universal gravitation $G \simeq 6.7 \times 10^{-8} cm^3/g \cdot s^2$ was determined, as well as the speed of light in vacuum $c \simeq 3 \times 10^{10} cm/s$ and Plancks's constant $\hbar \simeq 10^{-27} erg \cdot s$. All the measurements involve errors, but Quantum Mechanics introduces a particular type of inherent errors in the measurement process.

11.2 Black body

We do not know how the thermodinamic equilibrium sets in; we only believe that after a sufficiently long time it is realized, and we believe we can perceive it from the constancy of the thermodynamic quantities, like the temperature; then, we believe the thermodynamic equilibrium exists. In an internal cavity in a large body the electromagnetic radiation may be in equilibrium with the mattter of the body. Then, we have a black body; measuring the temperature we may infere the radiation energy. We need access to the black body, so we should practise a small aperture in it, which upsets the equilibrium, by emitting and absorbing radiation. We can send external radiation inside the cavity and measure the raise in the temperature; then we may have a very approximate measurement of the radiation energy flux. Or, we may extract a tiny amount of black-body radiation from a black body, filter it through a dispersion grating, or a prism, and measure its energy (by a black body); by means of the formula of the black-body radiation energy we may have access to Planck's constant.

11.3 Light

We may heat up gases and do interference with the light emitted by them when heated; we see that there exist characteristic spectra of gases, with some discrete wavelengths; many other wavelengths are absent in these spectra. From the characteristic spectra of gases Bohr's idea of quantum-mechanical states of the atoms appeared. By the black-body radiation theory, from the characteristic spectra of gases and from the photoelectric effect it appeared Einstein's idea of energy quanta (the photon of light) and energy quantization ($\varepsilon = h\nu$, where ε is the energy, ν is the frequency and $h \simeq 6 \times 10^{-27} erg \cdot s$ is Planck's constant).

11.4 Momentum

The variation $\delta\varepsilon = c\delta p$ in radiation energy, where p is momentum, implies, according to the energy quantization, $\delta\varepsilon = c\delta p = h\delta\nu$, hence the momentum quantization $p = h\nu/c = h/\lambda$, where λ is the radiation wavelength; radiation waves are particles (photons). Similarly, for a particle $\delta\varepsilon = v\delta p = h\delta\nu$ and $p = h/\lambda$, where $\lambda = vT$, $T = 1/\nu$ being the wave period; particles are waves, this was de Broglie's inference. Electrons diffracted by crystals (Davisson and Germer) confirmed the wave-like nature of the electrons. Similarly, the double-slit experiment confirmed the wave-like structure of the particles.

Radiation wavelengths can be measured by interference, hence we can infer the radiation momentum. What about particles, like electrons? We can heat up a piece of metal until vaporization, a local plasma of electrons and ions is formed (is it stable?), then apply a voltage bias and extract an electron beam through an aperture. In plasma (as in solids) the most energetic electrons (which are extracted) have wavelengths of the order of the mean separation distance of the ions; it follows that the transverse wavelengths of the electrons in the beam is of the order of the atomic dimensions (a few Ås), even the width of the beam is much larger (μm, mm); the beam has a filamentary structure, with a small dispersion, and with a flux $\simeq 10^{14} cm^{-2}$ at most (electron density in a solid $\simeq 10^{22} cm^{-3}$); the electrons are localized over transverse directions; usually the quasi-classical approximation may be applied for these electron beams; a similar situation holds for ion beams. The energy of the (charged) particles (and their momentum) can be measured by a voltage which stops them.

Many identical particles in a cavity may have wider wavefunctions, they are more delocalized (including the transverse directions), and their wavefunctions may overlap to a large extent.

11.5 Particle position

Waves are defined over distances larger than their wavelength (and in times longer than their oscillations period). A photon of green light has the wavelength $\lambda_{ph} \simeq 5 \times 10^3$Å, an electron with energy $1keV$ has the wavelength $\lambda_e \simeq 0.5$Å. We can collimate a light beam through an aperture larger than λ_{ph} and send it upon an electron; the electron will scatter the radiation in all directions, we can collect a beam of scattered radiation, pass it through a lens and determine its direction; the intersection of the incident and scattered light beams will give approximately the electron position; however, the focal region of the lens is larger than λ_{ph}, then we determine the electron position with uncertainty λ_{ph}; leaving aside that after collision the electron position changes (and leaving aside the lens aberration). This is the origin of Heisenberg's uncertainty relation $\Delta x \Delta p \simeq h$, where $\Delta p = h/\lambda$ is the momentum uncertainty and $\Delta x = \lambda$ is the position uncertainty.

11.6 Energy

An uncertainty in momentum implies an uncertainty in energy, $\Delta\mathcal{E} = v\Delta p = hv/\lambda = h/T$; but T is the minimum uncertainty in time, so $\Delta\mathcal{E}\Delta t \simeq h$; if the uncertainty in momentum is smaller, and the uncertainty in position is larger, then the uncertainty in time is larger and the energy is measured with a higher accuracy; a sharp energy is measured in an infinite time.

11.7 Angular momentum

An electron in atom may have an angular momentum L; then it has an orbital magnetic moment $\mu_0 = (e\hbar/2mc)L$, where e is the (negative) electron charge; it has also an intrinsic magnetic moment $\mu_s = (e\hbar/mc)s$, arising from the spin s. In a magnetic field $\mathbf{H}$, it acquires an energy $\mathcal{E} = -\boldsymbol{\mu}\mathbf{H}$; a force $\mathbf{F} = -grad\mathcal{E}$ appears, if the field is inhomogeneous, and the atoms beam is splitted, according to the total angular momentum; for free electrons, according to their 1/2-

spin; this way we may have access to the orbital angular momentum L (Stern-Gerlach experiment).

11.8 Collapse of the wavefunction

We can prepare electrons (atomic particles, in general) with large wavelengths, and then send them upon a sensitive screen (scintillator); they are delocalized over large distances, but are absorbed by localized atoms in the scintillator; their wavefunction becomes localized; this happens with a small, random probability. This is the collapse of the wavefunction. What makes the wave to shrink itself to atomic dimensions? It is a quantum-mechanical jump. There is no mechanism to describe the quantum jumps. There is no means to measure fully a wavefunction.

As far as the wavefunction is a wave of probability, it relates to our knowledge of the electron; does it relate to the electron too? Is the wavefunction about electron, or about our knowledge of the electron?

In the double-slit experiment we can measure the position of the electrons (by light) and each time we find the electron near a slit or other; without this measurement process, the electron (as an individual entity) goes through both slits, collapse upon one atom or other on the scintillator, and many collapsed electrons form a diffraction pattern on the screen; that means the electrons behave really, intrinsically, as waves; that wave would not be a subjective description of our knowledge. Still, rumour goes that Bohr used to say (approximately) that science is not about nature, it is about our knowledge of the nature.

The experimental evidence provided convincing paths of arriving at the quantum-mechanical theory; but, still, we would like to know more about why the Quantum Theory is the way it is. It is often said that we do not understand completely the Quantum Theory, in contrast to Mechanics, Statistical Physics or Electromagnetism. "The best that most of us can hope in physics is simply to misunderstand at a deeper level" (Pauli).

For instance, the absolute, ideal space requires a constant velocity $\mathbf{v}$ for a free particle; then it is reasonable to assume $d(m\mathbf{v})/dt = \mathbf{F}$ for a particle with mass m under the action of a force $\mathbf{F}$. The statistical equilibrium is reasonable for the highest disorder; electric charges and Lorentz force can be reasonably accepted and understood (though a danger of mistaking charges for fields and viceversa exists). But why probabilistic particles may have definite laws of motion (for unmeasurable quantities like the wavefunction)? The Quantum Mechanics endows the particles with a new quantity, the probability, represented by the wavefunction, which obeys a definite (Schrodinger's equation), which is the equation of energy conservation.

11.9 Measurement

One of the most general form of motion in space and time is described by the variation

$$\mathbf{k}d\mathbf{r} - \omega dt = d\Phi \ , \tag{11.1}$$

where $\mathbf{r}$ is the position vector, t denotes the time and $\mathbf{k}$ and ω are some constants; what is relevant here is $\mathbf{k}$ along $d\mathbf{r}$; if this $\mathbf{k}$ is such that $k = 2\pi/\lambda$ and $\omega = 2\pi/T$, then Φ is the phase of a wavefunction; λ is the wavelength and T is the period, $\mathbf{k}$ is the wavevector, ω is the frequency of a wave; any wavefunction can be expanded in a Fourier series of such "plane waves"; to get the connection with the Mechanics, we multiply the phase by Planck's constant $\hbar$, get the momentum $\mathbf{p} = \hbar\mathbf{k}$, the energy $E = \hbar\omega$, the mechanical action $S = \hbar\Phi$ and write the wavefunction as

$$\psi = e^{\frac{i}{\hbar}S} \ . \tag{11.2}$$

we can see that a wavefunction can get a mechanical interpretation and may serve to describe a general motion; moreover, the Φ is relativistically invariant. This sort of description of the motion is the Quantum Mechanics. We may see that we get the momentum and energy by applying the differential operators $-i\hbar\partial/\partial\mathbf{r}$ and $i\hbar\partial/\partial t$, respectively, to the wavefunction. We leave aside here the machinery of the Quantum Mechanics (including the probability, the superposition principle, the operators, matrices, eigenstates, commutation relations, etc) and focus on the fundamental restrictions implied by the wavefunction.

First, we note that a reproducible result of a measurement of the position x at a fixed moment of time is meaningless inside the wavelength;

similarly, a reproducible measurement of the time at a given position is meaningless inside a period; or, equally well, a measurement of the momentum can only be made with the maximal accuracy h/λ and a measurement of the energy can only be made with the maximal accuracy h/T. This follows more formally from the indeterminacy $\simeq \pi$ in the phase of any wavefunction, *i.e.*

$$\Delta p \Delta x \simeq h/2 \ , \ \ \Delta E \Delta t \simeq h/2 \ . \tag{11.3}$$

More formally, let us form

$$\left| -i\hbar\frac{\partial}{\partial x} - \lambda x \right|^2 \geq 0 \ , \tag{11.4}$$

or

$$-\hbar^2\frac{\partial^2}{\partial x^2} + i\hbar\lambda + \lambda^2 x^2 \geq 0 \ , \tag{11.5}$$

where λ is a parameter; this trinomial in λ is positive for

$$\Delta p \Delta x \geq \hbar/2 \tag{11.6}$$

(what we always measure is difference in positions, momenta, etc). A similar relation is obtained similarly for energy and time. These are Heisenberg's uncertainty relations. Moreover, since

$$h/2 \leq \Delta p \Delta x = h\frac{\Delta\lambda}{\lambda^2}\Delta x \leq h\Delta x/\lambda \ , \tag{11.7}$$

we get

$$\Delta x \geq \lambda/2 \ ; \tag{11.8}$$

and, since

$$h/2 \leq \Delta E \Delta t = h\frac{\Delta T}{T^2}\Delta t \leq h\Delta t/T \ , \tag{11.9}$$

we get

$$\Delta t \geq T/2 \ . \tag{11.10}$$

We can see that the maximal accuracy of any measurement in a wavefunction is $\lambda/2$ in position and $T/2$ in the time. Hence, by using the uncertainty relations, the maximal accuracy in the measurement of the momentum is h/λ and in energy is h/T. It follows that in using wavefunctions in Quantum Mechanics we cannot have information for distances shorter than the wavelength (at a given moment of time), neither for durations of time shorter than the wave period (at a given position).

These restrictions have important consequences for quantum relativity. The uncertainty by which we can measure the velocity of the interaction is of the order

$$\Delta v = \frac{\Delta x}{\Delta t} = \frac{\lambda}{T} = \frac{\omega}{k} = \frac{E}{p} = c\sqrt{1 + \frac{m^2c^2}{p^2}} \ , \qquad (11.11)$$

where m is the mass of the particle and c is the speed of light. We can see that the measurements in quantum relativity imply velocity higher than the speed of light, which means that the quantum relativity and the quantum theory of the relativistic fields are, basically, meaningless, when they are put in these terms; since they operate with entities which are not measurable, according to the principle of relativity. Since it is difficult to give up the principle of relativity, we have no other choice than to accept that the quantum treatment of the relativistic fields is, essentially, inappropriate, in terms of the usual quantum-mechanical treatment; we have no other choice but to admit that the position and the time in relativistic quantum field theory are not physical quantities, they are parameters.

In the rest frame the maximal accuracy of measurement of the position is $\Delta x \simeq \lambda/2 = h/2mc$, where $\lambda = h/mc$ is the Compton wavelength; the maximal accuracy of the measurement of the momentum is mc; the maximal accuracy of the measurement of the energy is mc^2; and the maximal accuracy of the measurement of the time is $\Delta t \simeq h/2mc^2$. It follows that we have no reliable method of predicting (describing) the production of particle-antiparticle pairs from vacuum, the vacuum polarization and breakdown, radiative corrections, mass and charge renormalization, etc, since we have no means of measuring these phenomena with sufficient accuracy.[1] The wavefunction for a relativistic particle with mc^2 for energy and mc for

[1]L. Landau and E. Lifshitz, *Course of Theoretical Physics*, vol. 4, V. B. Berestetskii, E. M. Lifshitz and L. P. Pitaevskii, *Quantum Electrodynamics*, Elsevier (2002); L. Landau, "On the Quantum Theory of Fields", in *Niels Bohr and the Development of Physics*, ed. W. Pauli, Pergamon Press (1955); W. Pauli, *General Principles of Quantum Mechanics*, Springer (1980); L. Landau and

momentum, $e^{-\frac{i}{\hbar}mc^2t+\frac{i}{\hbar}mcx} = e^{-\frac{i}{\hbar}mc(ct-x)}$, is meaningless in Quantum Mechanics for $c\Delta t \simeq \Delta x \simeq h/mc$; in any reference frame; on the other hand, for $c\Delta t$, $\Delta x \gg h/mc$ it describes a (quasi-) classical particle; the relativistic particles are (quasi-) classical particles, since all the durations of time and all the distances are much larger than the Compton period h/mc^2 and Compton wavelength h/mc. Similarly, if we view $e^{-\frac{i}{\hbar}p(ct-x)} = e^{-ik(ct-x)} = e^{-i2\pi(ct-x)/\lambda}$ as a radiation (photon) wavefunction, it describes a (quasi-) classical particle for times and distances much larger than λ/c and λ, respectively; and a quantum particle for times and distances larger, but comparable, with λ/c and λ, respectively.

We note that in the above considerations the position and the time are viewed as independent variables. So are the energy and the momentum, inasmuch as they are in fact differences of momenta and energies. This is specific to the non-relativistic Quantum Mechanics. In Relativity the situation is quite different. The energy and momentum are related by a definite rest energy mc^2, through $E = \sqrt{m^2c^4 + p^2c^2}$ for a free particle (bound states are inappropriate in full Relativity, and particles dressed with interaction preserve the relation). Similarly, the position and the time are related by the finite, maximal speed of light. Then, we have the absolute uncertainty relations $\Delta x > h/mc$, $\Delta p > mc$, $\Delta E > mc^2$, $\Delta t > h/mc^2$. For an electron, the Compton wavelength is $h/mc = 2 \times 10^{-10}cm$, the rest momentum is $mc = 3 \times 10^{-17}g \cdot cm/s$, the rest energy is $mc^2 = 9 \times 10^{-7}erg = 0.5MeV$ and the minimal time is $h/mc^2 = 6 \times 10^{-21}s$. For more massive particles, the Compton wavelength and the minimal time are even smaller; it follows that for all distances and times of interest the relativistic particles are in the quasi-classical limit. The quantum theory of relativistic particles gives up its usual terms, and transfers the quantum character to the fields, which describe quanta of momentum and energy, associated to particles.

Let us turn to the measurement process. Let

$$\psi = \sum_n a_n \psi_n \tag{11.12}$$

R. Peierls, "Erweiterung des Unbestimmtheitsprinzips fur die relativistische Quantentheorie", Z. Physik **69** 56 (1931); L. Landau and R. Peierls, "Quantenelektrodynamik im Konfigurationsraum", Z. Physik **62** 188 (1930).

be an expansion of ψ in a series of (orthogonal) wavefunctions ψ_n; then

$$\int d\mathbf{r}\psi_n^* \psi = a_n \; ; \qquad (11.13)$$

on the other hand,

$$\int d\mathbf{r} \left|\psi\right|^2 = \sum_n \left|a_n\right|^2 \int d\mathbf{r} \left|\psi_n\right|^2 \; ; \qquad (11.14)$$

if $\int d\mathbf{r} \left|\psi\right|^2$ is the probability of being in the state ψ in the volume of integration (with proper normalization), and $\int d\mathbf{r} \left|\psi_n\right|^2$ is the probability of being in the state ψ_n in the same volume, then $\left|a_n\right|^2$ is the probability that state ψ_n be included in state ψ in the volume of integration. It follows that we can take a_n as the amplitude of probability that state ψ_n be in state ψ; or, the extent to which the state ψ_n is included in state ψ. Therefore, the quantity

$$a = \int d\mathbf{r}\psi_1^* \psi_2 \qquad (11.15)$$

is an expression of a measurement process: its squared modulus indicates the amount to which the state ψ_1 is measured by state ψ_2. In a more descriptive sense, ψ_2 is the instrument (the apparatus) by which we measure the presence and the extent of ψ_1, and the process of measurement is the integration in equation (11.15). Obviously, these statements are valid for any generalized coordinate.

Let us assume that we try to gauge the state ψ_1 by its momentum; then we use

$$\psi_2 = \frac{1}{\sqrt{\Delta V}} e^{\frac{i}{\hbar}\mathbf{p}'\mathbf{r}} \; ; \qquad (11.16)$$

it is conceivable that our meter has several values of $\mathbf{p}'$ distributed within the range $\Delta\mathbf{p}'$ (around some value $\mathbf{p}'$), so we may use

$$\psi_2 = \frac{1}{\Delta\mathbf{p}'\sqrt{\Delta V}} \int d\mathbf{p}' e^{\frac{i}{\hbar}\mathbf{p}'\mathbf{r}} \; ; \qquad (11.17)$$

we get

$$a = \frac{1}{\Delta\mathbf{p}'\sqrt{\Delta V}} \int d\mathbf{p}' \int d\mathbf{r}\psi_1^* e^{\frac{i}{\hbar}\mathbf{p}'\mathbf{r}} \; ; \qquad (11.18)$$

let us assume that ψ_1 has a certain $\mathbf{p}$ to some extent $a_{\mathbf{p}}$,

$$\psi_1 = \frac{1}{\sqrt{\Delta V}} a_{\mathbf{p}} e^{\frac{i}{\hbar}\mathbf{pr}} \; ; \tag{11.19}$$

then,

$$a = \frac{h^3}{\Delta \mathbf{p}' \Delta V} \int d\mathbf{p}' a_{\mathbf{p}}^* \delta(\mathbf{p} - \mathbf{p}') = \frac{h^3}{\Delta \mathbf{p} \Delta V} a_{\mathbf{p}}^* \; . \tag{11.20}$$

If $\Delta\mathbf{p}\Delta V$ is very large, *i.e.* the accuracy of the measurement is very low, and our apparatus ψ_2 is a quantum-mechanical apparatus, then a is extremely small, and we have no definite result; our measurement process fails. On the contrary, if the apparatus is classical, *i.e.* $\Delta\mathbf{p}\Delta V \simeq h^3$, then $a \simeq a_{\mathbf{p}}$, which means that, indeed, we succeed to measure the state ψ_1 defined by its momentum. However, it is worth noting that the quantum-mechanical character of the product $\Delta\mathbf{p}\Delta V$ is valid for finite $\Delta\mathbf{p}$ and ΔV and large values of the product $\Delta\mathbf{p}\Delta V$.

Similarly, we can use an apparatus

$$\psi_2 = \frac{1}{\Delta E'} \int dE' e^{-\frac{i}{\hbar}E' t} \phi_2(\mathbf{r}) \tag{11.21}$$

for measuring a state

$$\psi_1 = e^{-\frac{i}{\hbar}Et} \phi_1(\mathbf{r}) \tag{11.22}$$

defined by its energy E; it is conceivable that the measuring process takes a finite time Δt, during which we collect the results; we get

$$a = \frac{1}{\Delta E' \Delta t} \int dt \int dE' e^{\frac{i}{\hbar}(E-E')t} \int d\mathbf{r} \phi_1^*(\mathbf{r}) \phi_2(\mathbf{r}) \; , \tag{11.23}$$

or, leaving aside the spatial factor,

$$a = \frac{h}{\Delta E' \Delta t} \int dE' \delta(E - E') = \frac{h}{\Delta E \Delta t} \; ; \tag{11.24}$$

if the apparatus is quantum-mechanical, there will be large uncertainties in energy and time, and the measurement process fails (it gives a vanishingly small, indefinite result); if the apparatus is classical, the uncertainties are of the order $\Delta E \Delta t \simeq h$, and we get a definite result

a; which shows to what extent a state defined by its energy is present in the measured state ψ_1.

Let us go back to the typical energy, or momentum wavefunctions of the form of the plane wave, $e^{-\frac{i}{\hbar}Et}$ or $e^{\frac{i}{\hbar}\mathbf{pr}}$. If we try to consider such wavefunctions as basic elements of reality, they are useless, as partly useless are their consequences like the quantum-mechanical energy levels, for instance. This is so because energy or momentum are not determined for finite intervals of time or finite regions of space, in the sense of the uncertainty relations. Consequently, the energy and the momentum are not conserved, which means the loss of scientific objectivity (reproducibility, control of errors, etc). This can also be seen in the process of measurement, where factors like $e^{\frac{i}{\hbar}(E-E^{'})t}$, or integrals over finite durations of time of such factors, are far away of $\delta(E-E^{'})$ (and similarly for the momentum). A typical measurement, for instance, is to send a monochromatic light wave upon a collection of atoms; we shall see that the absorption law $\hbar\omega = \Delta E$ (for absorbing one photon), where ω is the radiation frequency and ΔE the difference between two atomic energy levels, is very scattered; and only after a long time ($t \gg 1/\hbar\omega$), we get a well-defined spectral peak. In the uncertainty relation $\Delta E \Delta t = nh$, for some large n, we may have $\Delta t = nh/\delta E$, such that $\Delta E = \delta E$, where δE is arbitrarily small; then, we can have a sharp measurement of the energy. It follows that we can still use a quantum-mechanical apparatus (an atom) for the measurement process, at the price of waiting a long time. However, for a long time, the wavefunction $\sim e^{-\frac{i}{\hbar}Et}$ is in its quasi-classical condition, since we have many oscillations. It follows that the quantum-mechanical concept has a meaning only in relation to the classical concepts; of course, as long as we accept as positive knowlege only what we are able to measure with a controllable accuracy; which is unavoidable for the scientific meaning.

12 References

1. P. A. M. Dirac, *The Principles of Quantum Mechanics*, Clarendon, Oxford (1947).
2. W. Pauli, *General Principles of Quantum Mechanics*, Springer, Berlin (1980).
3. J. Frenkel, *Wave Mechanics*, Clarendon, Oxford (1934).
4. A. Sommerfeld, *Atombau und Spektrallinien*, Vieweg&Son, Braunschweig (1939).
5. M. Born and P. Jordan, *Elementare Quantenmechanik*, Springer (1930).
6. H. A. Kramers, *Quantum Mechanics*, Dover, NY (1964).
7. G. Wentzel, *Einfuhrung in die Quantentheorie der Wellenfelder*, Deuticke, Vienna (1943).
8. N. F. Mott and H. S. W. Massey, *The Theory of Atomic Collisions*, Clarendon, Oxford (1949).
9. E. U. Condon and G. H. Shortley, *Theory of Atomic Spectra*, Cambridge University Press, Cambridge (1951).
10. W. Heitler, *The Quantum Theory of Radiation*, Clarendon, Oxford (1954).
11. H. A. Bethe and E. E. Salpeter, *Quantum Mechanics of One- and Two-Electron Atoms*, Springer, Berlin (1957).
12. D. Bohm, *Quantum Theory*, Dover, NY (1951).
13. L. I. Schiff, *Quantum Mechanics*, McGraw-Hill, NY (1955).
14. L. Landau and E.Lifshitz, *Course of Theoretical Physics*, vol. 3, *Quantum Mechanics (Non-relativistic Theory)*, Elsevier, Oxford (2005).
15. K. Gottfried and Tung-Mow Yan, *Quantum Mechanics: Fundamentals*, Springer, NY (2003).

16. N. F. Ramsey, *Nuclear Moments*, Wiley, NY (1953).
17. A. R. Edmonds, *Angular Momentum in Quantum Mechanics*, Princeton University Press, Princeton (1957).
18. I. S. Gradshteyn and I. M. Ryzhik, *Table of Integrals, Series and Products*, eds. A. Jeffrey and D. Zwillinger, Academic Press (2000).
19. *Handbook of Mathematical Functions*, eds. M. Abramowitz and I. A. Stegun, Nat. Bur. Stand. (1972).
20. V. Galitski, B. Karnakov, V.Kogan and V. Galitski, Jr, *Exploring Quantum Mechanics (Collection of Problems)*, Oxford University Press (2013).
21. D. ter Haar (Goldman Krivchenkov, Kogan and Galitski, *Problems in Quantum Mechanics*), *Selected Problems in Quantum Mechanics*, Infosearch Ltd London (1964).

Index